AIRO Springer Series

Volume 2

The **AIRO Springer Series** focuses on the relevance of operations research (OR) in the scientific world and in real life applications.

The series publishes contributed volumes, lectures notes, and monographs in English language resulting from workshops, conferences, courses, schools, seminars, and research activities carried out by AIRO, Associazione Italiana di Ricerca Operativa - Optimization and Decision Sciences: http://www.airo.org/index.php/it/.

The books in the series will discuss recent results and analyze new trends focusing on the following areas: Optimization and Operation Research, including Continuous, Discrete and Network Optimization, and related industrial and territorial applications. Interdisciplinary contributions, showing a fruitful collaboration of scientists with researchers from other fields to address complex applications, are welcome. The series is aimed at providing useful reference material to students, academic and industrial researchers at an international level.

More information about this series at http://www.springer.com/series/15947

Mauro Dell'Amico · Manlio Gaudioso ·
Giuseppe Stecca
Editors

A View of Operations Research Applications in Italy, 2018

AIRO
ASSOCIAZIONE ITALIANA DI RICERCA OPERATIVA
OPTIMIZATION AND DECISION SCIENCE

Springer

Editors
Mauro Dell'Amico
DISMI - Department of Sciences
and Methods for Engineering
University of Modena and Reggio Emilia
Reggio Emilia, Italy

Manlio Gaudioso
DIMES
University of Calabria
Arcavacata di Rende, Italy

Giuseppe Stecca
CNR - IASI Institute for Systems Analysis
and Computer Science
National Research Council
Roma, Italy

ISSN 2523-7047 ISSN 2523-7055 (electronic)
AIRO Springer Series
ISBN 978-3-030-25841-2 ISBN 978-3-030-25842-9 (eBook)
https://doi.org/10.1007/978-3-030-25842-9

This Springer imprint is published by the registered company Springer Nature Switzerland AG
The registered company address is: Gewerbestrasse 11, 6330 Cham, Switzerland

Preface

In 2017, the Italian Operations Research Society (AIRO) called upon its members to contribute detailed information about their activities in promoting the use of quantitative techniques, and in particular the operations research ones, in society and industry.

The collected information was condensed into a report presenting 74 case studies highlighting the real pervasiveness of operations research techniques. The booklet was issued for the annual AIRO conference ODS2017, held in Sorrento, and the areas covered encompassed both the public and the private sectors, involving Logistics and Transportation, Production, Services and Society, ICT, and Energy and Environment.

It was particularly exciting to observe the very productive and fruitful cooperation between the operations research people, from both academic institutions and spinoff companies, and domain experts from the diverse application frameworks.

AIRO decided that the experience gathered by so many research groups, working all over Italy and sometimes even abroad, deserved an appropriate and prestigious outlet that would assist in encouraging new and extended applications.

The best opportunity was provided by the recently established cooperation between AIRO and Springer, and a call for papers, primarily but not exclusively directed at contributors to the report, was launched at the beginning of 2018.

The present volume is the outcome of the above call. It contains 16 papers which offer an excellent view on the application of operations research in several fields by the use of effective approaches. All the published papers are based on successful collaborations with companies which confirmed the positive impact of the research on their processes (in some papers, company representatives are also co-authors).

Papers are clustered according to the application areas that they cover. In particular, papers 1–4 describe several applications in Logistics, 5 and 6 are Product Design oriented, 7 and 8 are about Production Planning and Scheduling, 9–11 are related to Social Applications, and, finally, 12–16 deal with Transportation problems with respect to both freight and people.

It is anticipated that the book will attract the interest of research scientists as well as professionals, and its use for didactic purposes is also possible.

We wish to thank all colleagues who have contributed their papers, Professor Daniele Vigo, President of AIRO, for his enthusiastic support, and Springer for having given us such an excellent opportunity.

Reggio Emilia, Italy — Mauro Dell'Amico
Arcavacata di Rende, Italy — Manlio Gaudioso
Rome, Italy — Giuseppe Stecca
May 2019

Contents

Production Planning and Scheduling

Social Applications

Transportation

Editors and Contributors

About the Editors

Mauro Dell'Amico is Full Professor of Operations Research at the University of Modena and Reggio Emilia. His main research interests focus on combinatorial optimization as primarily applied to mobility, logistics, transportation, and production planning and scheduling. He is the author of *Assignment Problems* (SIAM 2012) and almost 70 papers on discrete optimization and related areas. He is a member of the board of the Italian Association for Operations Research (AIRO) and President of the Interuniversity Consortium for Optimization and Operations Research. He has participated as principal investigator in many international funded research projects.

Manlio Gaudioso holds a "Laurea" degree in Electrical Engineering from Università di Napoli. Since 1994, he has been Full Professor of Operations Research at Università della Calabria. His research interests include nonsmooth optimization, integer programming, graph optimization, logistic chain optimization, and classification methods in machine learning. He is currently Associate Editor of the journal "Optimization" and of "Vestnik of Saint Petersburg University". He plays drums in the Italian Jazz Band "Ars Brevis".

Giuseppe Stecca is a Research Scientist at the Institute of Systems Analysis and Computer Science "Antonio Ruberti" of the Italian National Research Council (CNR-IASI). He holds the Chair of Supply Chain Management at the University of Rome "Tor Vergata", where he also teaches Operations Research. He is a member of the board of the Italian Association for Operations Research (AIRO). His main research interests are related to the optimization of sustainable production and logistic systems. He works actively in research projects and also as an evaluator for the Italian Ministry of Economic Development in the area of logistics and industry 4.0.

Contributors

Daniela Ambrosino DIEC, Universitá di Genova, Genova, Italy

Claudio Arbib DISIM, University of L'Aquila, Abruzzo, Italy

Veronica Asta CIELI, Universitá di Genova, Genova, Italy

Tiziano Bacci Istituto di Analisi ed Informatica "A. Ruberti" del CNR, Rome, Italy

Federico Bartoli Costa Crociere Spa, Genova, Italy

Luca Bertazzi Department of Economics and Management, University of Brescia, Brescia, Italy

Martin Bué Inria Lille-Nord Europe, Lille, France

Roberto Calbi AutoTerminal Gioia Tauro S.p.A, Gioia Tauro, Italy

Samuela Carosi Department of OR Development, Lucca, Italy

Diego Cattaruzza University Lille, CNRS, Centrale Lille, Inria, UMR 9189 - CRIStAL - Centre de Recherche en Informatique Signal et Automatique de Lille, Lille, France

Claudio Ciancio Department of Mechanical, Energy and Management Engineering, University of Calabria, Rende, CS, Italy

Stefano Conte AreSoft S.r.l., Rome, Italy

Andrea Credo Dipartimento di Ingegneria Industriale e dell'Informazione e di Economia, Università degli studi dell'Aquila, L'Aquila, Italy

Andrea Cristofari Dipartimento di Matematica, Università degli studi di Padova, Padua, Italy

Frans Cruijssen ArgusI, Breda, The Netherlands

Mauro Dell'Amico DISMI, Università di Modena e Reggio Emilia, Reggio Emilia, Italy

Gianmaria De Tommasi DIETI, Università di Napoli Federico II, Napoli, Italy; Consorzio CREATE, Napoli, Italy

Luigi Di Puglia Pugliese DIMEG, Universitá della Calabria, Rende (CS), Italy

Edoardo Fadda DAUIN, Politecnico di Torino, Torino, Italy

Antonio Frangioni Dipartimento di Informatica, Università di Pisa, Pisa, Italy

Guenther Fuellerer STRAsco Team, Zentrale Technik, Strabag AG, Vienna, Austria

Laura Galli Dipartimento di Informatica, Università di Pisa, Pisa, Italy

M. Gentili Industrial Engineering Department, University of Louisville, Louisville, KY, USA

E. Gerber Industrial Engineering Department, University of Louisville, Louisville, KY, USA

Giovanni Giallombardo Dipartimento di Ingegneria Informatica, Modellistica, Elettronica e Sistemistica (DIMES), Università della Calabria, Rende, Italy

Leopoldo Girardi Department of OR Development, Lucca, Italy

K. Gue Industrial Engineering Department, University of Louisville, Louisville, KY, USA

Francesca Guerriero DIMEG, Universitá della Calabria, Rende (CS), Italy

Gerhard Hoefinger STRAsco Team, Zentrale Technik, Strabag AG, Vienna, Austria

Demetrio Laganá Department of Mechanical, Energy and Management Engineering, University of Calabria, Rende, CS, Italy

Lorenzo Lampariello Department of Business Studies, Roma Tre University, Rome, Italy

Stefano Lucidi Dipartimento di Ingegneria Informatica, Automatica e Gestionale, Sapienza Università di Roma, Rome, Italy

Francesca Maggioni Department of Management, Economics and Quantitative Methods, University of Bergamo, Bergamo, Italy

Andrea Manno Dipartimento di Elettronica, Informazione e Bioingegneria, Politecnico di Milano, Milan, Italy

Domenico Matera AreSoft S.r.l., Rome, Italy

Sara Mattia Istituto di Analisi ed Informatica "A. Ruberti" del CNR, Rome, Italy

Giovanna Miglionico Dipartimento di Ingegneria Informatica, Modellistica, Elettronica e Sistemistica (DIMES), Università della Calabria, Rende, Italy

Mahyar T. Moghaddam DISIM, University of L'Aquila, Abruzzo, Italy

Henry Muccini DISIM, University of L'Aquila, Abruzzo, Italy

Roberto Musmanno Department of Mechanical, Energy and Management Engineering, University of Calabria, Rende, CS, Italy

André C. Neto Fusion for Energy, Barcelona, Spain

Lucia Nitoglia FATER S.p.A., Pescara, Italy

Stefano Novellani DISMI, Università di Modena e Reggio Emilia, Reggio Emilia, Italy

Maxime Ogier University Lille, CNRS, Centrale Lille, Inria, UMR 9189 - CRIStAL - Centre de Recherche en Informatique Signal et Automatique de Lille, Lille, France

Tiziano Parriani Optit srl, Imola, Bologna, Italy

Sergio Pelle FATER S.p.A., Pescara, Italy

Guido Perboli DAUIN, Politecnico di Torino, Torino, Italy

Matteo Pozzi Optit srl, Imola, Bologna, Italy

Francesco Rinaldi Dipartimento di Matematica, Università degli studi di Padova, Padua, Italy

Francesco Romito Dipartimento di Ingegneria Informatica, Automatica e Gestionale, Sapienza Università di Roma, Rome, Italy

Simone Sagratella Department of Computer, Control, and Management Engineering Antonio Ruberti at Sapienza University of Rome, Rome, Italy

Marco Santececca Dipartimento di Ingegneria Industriale e dell'Informazione e di Economia, Università degli studi dell'Aquila, L'Aquila, Italy

Francesco Santoro Department of Mechanical, Energy and Management Engineering, University of Calabria, Rende, CS, Italy

Frédéric Semet University Lille, CNRS, Centrale Lille, Inria, UMR 9189 - CRIStAL - Centre de Recherche en Informatique Signal et Automatique de Lille, Lille, France

Antonio Sforza DIETI, Università di Napoli Federico II, Napoli, Italy

Claudio Sterle DIETI, Università di Napoli Federico II, Napoli, Italy;
IASI-CNR, Rome, Italy

Roberto Tadei DAUIN, Politecnico di Torino, Torino, Italy

Giuliano Vallese Department of OR Development, Lucca, Italy

Paolo Ventura Istituto di Analisi ed Informatica "A. Ruberti" del CNR, Rome, Italy

Daniele Vigo Optit srl and DEI, University of Bologna, Bologna, Italy

Marco Villani Dipartimento di Ingegneria Industriale e dell'Informazione e di Economia, Università degli studi dell'Aquila, L'Aquila, Italy

Logistics

A Two-Phase Approach for an Integrated Order Batching and Picker Routing Problem

Martin Bué, Diego Cattaruzza, Maxime Ogier and Frédéric Semet

Abstract This article addresses an integrated warehouse order picking problem. The company HappyChic is specialized in men's ready-to-wear. A central warehouse is dedicated to supplying, every day, the shops of one brand. We focus on the picking area of this warehouse which relies on human picking system. For each picking wave (period of a working day), a set of customer orders has to be prepared. An order is a set of product references, with quantities, i.e., the numbers of items required. The problem consists in jointly deciding: (1) the division of orders into several boxes, respecting weight and size constraints; (2) the batching of boxes into trolleys, that implicitly defines the routing into the picking area. The objective function aims to minimize the total distance. To deal with the large size instances of HappyChic in short computation times, we design a heuristic method based on the split and dynamic programming paradigms. The results are very convincing: the total covered distance decreases by more than 20%. Moreover, we propose an adaptation of the algorithm to prepare homogeneous boxes with respect to classes of products. The logistic department of HappyChic is convinced by results obtained in this research work, and the warehouse management system is currently being updated in order to integrate the proposed algorithm.

Keywords Warehouse management · Order batching · Picker routing

M. Bué
Inria Lille-Nord Europe, Lille, France
e-mail: martin.bue@inria.fr

D. Cattaruzza (✉) · M. Ogier · F. Semet
Univ. Lille, CNRS, Centrale Lille, Inria, UMR 9189 - CRIStAL - Centre de Recherche en Informatique Signal et Automatique de Lille, 59000 Lille, France
e-mail: diego.cattaruzza@centralelille.fr

M. Ogier
e-mail: maxime.ogier@centralelille.fr

F. Semet
e-mail: frederic.semet@centralelille.fr

M. Dell'Amico et al. (eds.), *A View of Operations Research Applications in Italy, 2018*, AIRO Springer Series 2, https://doi.org/10.1007/978-3-030-25842-9_1

1 Introduction

In this paper, we focus on the warehouse management problem faced by our industrial partner HappyChic.[1] HappyChic is a French firm specialized in men's ready-to-wear. HappyChic gathers three leading brands: Jules, Brice, and Bizzbee. The logistic department of HappyChic is in charge of the three warehouses (one for each brand), and all the related logistics activities. In this paper, we consider operations in the warehouse dedicated to clothing products of the brand Jules, located in the North of France, close to the city of Lille. The warehouse supplies the shops of Jules in all France and some shops abroad. Thus, the usual customers to satisfy are the shops to which new clothing products are shipped.

The warehouse under consideration consists in four main zones: (1) the *reception zone* where products delivered by suppliers are received; (2) the *storage zone* where pallets of products are stored in racks; (3) the *picking zone* where pickers pick items from cartons and prepare boxes to be sent to customers; (4) the *delivery zone* where trucks are loaded with boxes shipped to customers.

The picking operations are done manually by pickers that push trolleys with several boxes. These boxes need to be filled with items required by customers.

Each working day a set of demands has to be prepared. The demands are associated with customers, i.e. shops located throughout France and abroad. Based on the sales, the shop demands are computed during the night. Demands usually consist of several items which may not be put into a single box. The demand of each customer has then to be split in different boxes and these boxes have to be assigned to one or several trolleys. Clearly one box must contain only products associated with a single customer. On the other hand, boxes on the same trolley can be for different customers.

In this paper, we focus on the optimization of the operations in the picking zone. The problem under study consists in optimizing the productivity of the pickers by simultaneously determining: (1) the products that constitute each box to be delivered to customers; and (2) the batch of boxes to constitute a picking tour.

2 Literature Review

Warehouse management involves different optimization problems such as conceiving picking tours, batching orders, storage assignment, layout design, and zoning (see, for example, [2]). Decisions to optimize productivity in a warehouse can have different horizons and be classified as strategic, tactical and operational. Strategic decisions determine the layout of the warehouse, the disposition of each zone (receiving, storage, picking and delivery) with respect to the others, the decisions that involve the storage policy as well as picking policy (automated versus human). Tactical decisions can involve the location of products based on their forecast demand in the storage and picking zone. At the operational level, storage and picking tours need

[1] http://www.happychicgroup.com.

to be efficiently determined and coordinated. Storage tours move products from the storage to the picking zone, while picking tours aim at collecting items to satisfy the demand of customers [2, 5].

Picking operations can be classified based on the automatism introduced in the process. As in [5] five classes can be identified: picker-to-parts, parts-to-picker, pick-to-box, pick-and-sort, and completely automated picking. In the first category, pickers move around the warehouse to pick items while in the parts-to-picker class, automated devices bring loads to pickers that are in charge of picking the right quantity required by the order under consideration. In pick-to-box systems, pickers are assigned to different zones and boxes containing customer orders move using a conveyor through different zones to be filled. In pick-to-sort systems, orders are batched and pickers collect items of a certain product to satisfy orders in the whole batch. A conveyor brings all the items to the sorting area where items are assigned to orders. Finally, humans do not intervene in full automated picking systems.

Picking systems where the human presence is necessary for operations still involve the majority of the warehouses [2]. Recent studies [3, 5] show how the cost of the automation would be too high to be profitable in a short or mid-term horizons. For these reasons and since in the problem under study picking operations can be classified as picker-to-parts, we concentrate the review on papers related to this area.

In the HappyChic case, our interest is focused on the operational decisions for picking operations. It concerns order batching and picker routing problems. Usually, order batching is performed when the size of orders is smaller than the capacity of the trolley used to collect the items. Order batching consists in grouping a set of orders into picking tours to minimize the total distance traveled by pickers. This problem is NP-hard [4]. Given a set of locations to visit, the picker routing problem aims to decide the sequence of locations to visit to minimize the traveled distance. These two problems have been studied extensively as separate problems [2]. Some recent works have addressed the joint order batching and picker routing problem. Valle et al. [8] propose a branch-and-cut algorithm based on a mixed integer programming formulation, and [7] proposed an iterated local search approach.

In the HappyChic case, we face a specific batching problem. Indeed, the size of an order is usually larger than the capacity of a trolley. Hence, the batching problem has two levels. First, items of a given order have to be batched into boxes. Second, boxes have to be assigned to trolleys. It is noteworthy that these two batching levels are mutually dependent.

3 Problem Definition

In this section, we formally introduce the problem and the notation used in the paper. All along the following sections, we suppose that each working day is divided into T periods. A period is associated with a picking wave. We indicate by $\mathscr{T} = \{1, \ldots, T\}$ the period set. The warehouse considered contains a set of products $\mathscr{P}$. With each product $p \in \mathscr{P}$ is associated the volume $v(p)$ and the weight $w(p)$ of a single item.

Each product p is also associated with a single location $l(p)$ in the picking zone. This assumption is satisfied for the HappyChic case since if a product is located in several locations, the locations are emptied based on a FIFO rule: first demands use the first location, and when this location is emptied, the second one is used. More precisely, let us detail this FIFO rule on a small example. Let us consider a product p located in two locations l_1 and l_2, such that location l_1 has to be emptied before l_2 according to the FIFO rule. We denote by Q_1 and Q_2 the available quantities of the product in locations l_1 and l_2 respectively. Then, instead of considering a single product p, we will consider two products p_1 and p_2 such that $l(p_1) = l_1$ and $l(p_2) = l_2$. These products have the same volume and weight. In the FIFO rule, demands are considered in a certain order. If the first demand requires $q_1 < Q_1$ units of p, then we consider q_1 units of product p_1 for this demand. Then location l_1 contains $Q'_1 = Q_1 - q_1$ available units of p_1. If the second demand requires q_2 units of p, such that $q_1 + q_2 > Q_1$, then we consider Q'_1 units of product p_1 and $q_2 - Q'_1$ units of p_2 for this demand. For the following demands of product p, only product p_2 will be considered since there is no more available quantity in location l_1. This FIFO rule is applied by HappyChic before the picking process and questioning this practice is not in the scope of the present work. So, given a demand, we can consider that there is no choice on the location to pick a product. Hence if a product is located in several locations, different products can be considered with the same volume and weight, and with only one unique location for each product.

3.1 Warehouse Description

The warehouse under consideration is depicted in Fig. 1 that represents the four zones dedicated to reception, storage, picking and delivery.

Trucks carrying products from the suppliers arrive at the reception zone of the warehouse. Here dedicated employees unload the trucks, and pallets of cartons are prepared to be stored. Then, employees driving forklifts move cartons from the reception zone into the racks of the storage zone. Other employees are dedicated to supplying the picking zone. They pick up cartons from pallets in the storage zone and bring them, using forklifts, into the picking zone. They are allowed only to go through cross-aisles 1, 3 and 5 and to leave cartons at the top or the bottom of the different aisles (see solid gray lines in Fig. 1).

All along the article, we make the distinction between *carton* and *box*. A carton contains items of a single product. Cartons are received by the warehouse on pallets and stored in the storage zone. They are then moved to the picking zone. There, items are picked from the cartons and put into boxes by dedicated employees called *pickers*. Thus, a box contains customer demands and possibly items of different products.

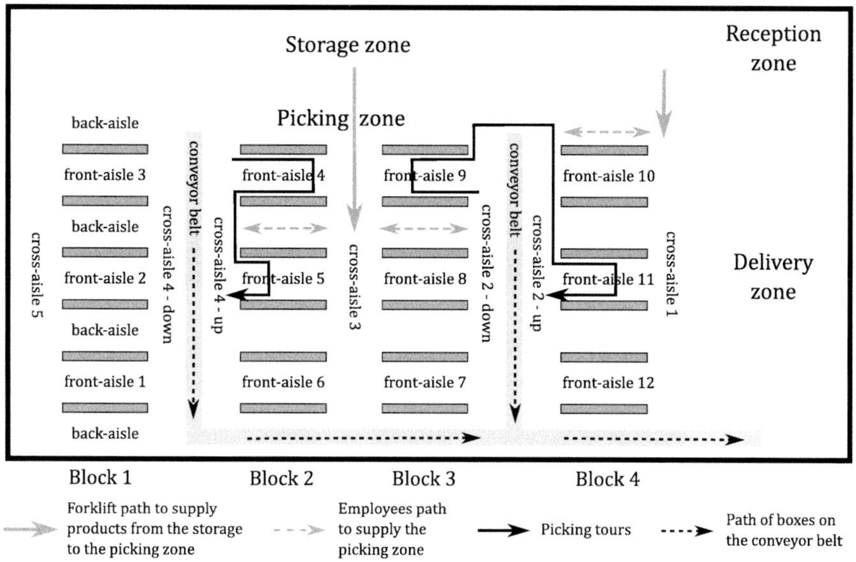

Fig. 1 The warehouse

3.2 *Picking Zone*

The picking zone of the warehouse consists of four blocks (see Fig. 1). Let us call these blocks, block 1 to 4. Each block contains different aisles having racks on one or both sides. Contrarily to standard cases presented in the literature (see for example [7]), we distinguish the two sides of the racks: the front and the back side. Cartons are inserted into racks from the back side and products are picked from the front side. Racks are situated in such a way that a front (resp. back) side always faces another front (resp. back) side of another rack (except, of course, for the first and the last rack). This placement defines a front-aisle and a back-aisle (see Fig. 2).

Back-aisles are used by employees to bring cartons from the top or the bottom of an aisle to their exact location (see dotted gray lines in Fig. 1). Only front-aisles are involved in the picking operations and are used by pickers to collect products from cartons.

The cross-aisles in the picking zone are not all walkable by pickers. For security reasons, cross-aisles 1, 3 and 5 are dedicated to forklifts that supply the picking zone from the storage zone. Pickers are allowed to walk only across cross-aisles 2 and 4. Moreover, these cross-aisles are equipped with a conveyor belt that automatically brings boxes prepared by pickers from the picking zone to the delivery zone. The conveyor belt divides cross-aisles 2 and 4 into two sub-cross-aisles. Let us use the words *up* and *down* to distinguish between these two sub-cross-aisles.

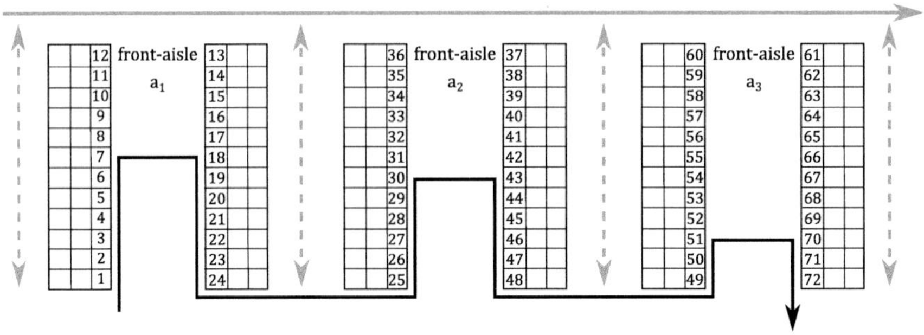

Fig. 2 Front-aisles and back-aisles. Legend as in Fig. 1. Locations are numbered following the walking order

Pickers walk around the warehouse pushing a trolley that contains up to f boxes. Due to the particular warehouse layout, the pickers cannot cross the cross-aisle 3. As a consequence, picking tours are limited to blocks 1 and 2 or to blocks 3 and 4.

Moreover, all the pickers follow the same circular path orientation and always walk following a clockwise walkway. This is due to space limitations in the aisles, and also facilitates picking operations since all routes are in the same direction. Thus, in a given route, the location names follow a topological order. We call this particular policy the *walking policy*. It is formally defined as follows.

First, aisles of blocks 1 and 2 and aisles of blocks 3 and 4 are ordered starting from the bottommost aisle of blocks 1 and 3 to the bottommost aisle of blocks 2 and 4 following the clockwise sense. Let $\mathscr{A}$ be the set of front-aisles. Given two front-aisles $a_1, a_2 \in \mathscr{A}$, aisle a_1 precedes aisle a_2 and we write $a_1 \prec a_2$, if a_1 comes before a_2 in this given order.

The set of all possible locations in the aisles of the picking zone is indicated by $\mathscr{L}$. A particular order is imposed on locations. Let us define the function $a : \mathscr{L} \rightarrow \mathscr{A}$ that associates each location with its aisle. Given two locations l_1 and l_2 such that $a(l_1) \neq a(l_2)$, we say that l_1 precedes l_2, and we indicate it by $l_1 \prec l_2$, if $a(l_1) \prec a(l_2)$. On the other hand, if $a(l_1) = a(l_2)$, we define the following rule. Entering a front-aisle from the cross-aisle, locations on the left are ordered from the beginning to the end of the front-aisle, while locations on the right are ordered on the opposite way. Thus, when $a(l_1) = a(l_2)$, we write $l_1 \prec l_2$ if:

- l_1 is on the left-side of the aisle, while l_2 on the right;
- l_1 and l_2 are both on the left-side and l_1 is nearer to the beginning of the aisle than l_2;
- l_1 and l_2 are both on the right-side and l_1 is nearer to the end of the aisle than l_2.

This order is called walking order or walking policy. An example is given in Fig. 2, where the numbers on each location follow the walking order. This walking policy corresponds to the return policy [2] where a picker enters and leaves each aisle from the same end.

Given two locations i and j, distances $d_{i,j}$ and $d_{j,i}$ are determined. They represent the distance of the path that goes from one location to the other following the walking order. Due to logistic constraints in the warehouse, it can be impossible to reach location j from location i (and/or vice versa). In this case, the distance $d_{i,j}$ is infinite. Location 0 represents a location where a tour starts and ends. Picking tours start and end at any place along the conveyor belt. This can be approximated by the entry point or exit point of an aisle. Hence location 0 is a fictive point that represents entrance or exit of an aisle. We then consider distance $d_{0,l}$ as the distance from the entrance point of aisle $a(l)$ to location l, and distance $d_{l,0}$ as the distance from location l to the exit point of aisle $a(l)$.

At a given period $t \in \mathcal{T}$, each location contains items of the same product. For each period t, we assume that the storage level in the picking zone is sufficient to satisfy the total demand.

3.3 Pickers

A team of employees is in charge of collecting items from the cartons in the picking zone and to fill the boxes that will be delivered to shops. These employees are called pickers. It is supposed that the number of pickers is large enough to ensure the preparation of the total demand to be prepared within a given period.

A picker walks around the warehouse respecting the walking policy. The picker pushes a trolley with up to f boxes available. He/she picks products from cartons and put them into the boxes. Once the all required items are picked, he/she walks to the conveyor belt on which he/she places the boxes. He/she then starts another picking tour with a trolley containing empty boxes. Then, the picker is allowed to walk against the walking policy to locate him/herself at the beginning of the next tour. Note that, due to the particular walking policy considered in the warehouse, the set of locations to visit to pick the items directly implies the path that a picker needs to follow to collect the items.

3.4 Demands

A set of demands $\mathcal{D}$ is to be prepared during a working day. The set $\mathcal{D}$ is partitioned in T subsets/periods, in which the working day is divided. Subset $\mathcal{D}_t$, $t \in \mathcal{T}$, contains all the demands that need to be prepared during period t.

For each demand $d \in \mathcal{D}$, a set of $n(d)$ products $\mathcal{P}_d = \{p_1^d, \ldots, p_{n(d)}^d\} \subseteq \mathcal{P}$ needs to be collected. A quantity q_j^d is associated with each product p_j^d in demand d. This corresponds to items of the same product, situated at the same location $l(p_j^d)$, and needed to prepare d.

Whenever useful for the presentation of this work when there is no ambiguity about the index of demand d, notations are lightened as follows. Given a demand d, the set of products is written $\{p_1, \ldots, p_n\}$. Each product p_j is associated with a quantity denoted as q_j, and a location l_j.

3.5 Boxes

Boxes to be delivered to shops must respect a maximal volume v_{box}^{max} and a maximal weight w_{box}. Moreover, each box must satisfy a minimal volume v_{box}^{min} of products. A box, that is not filled at least as much as the minimal volume, can be crushed by other boxes during transportation. Filling up a box increases its strength. Note that it may happen that a demand is too small to ensure the minimal volume capacity constraint. It is then the decision of the managers of the warehouse to decide if this demand is prepared (with violation of the capacity constraint) or delayed to another day. Here, we minimize the number of boxes that violated the minimal volume constraint and consider that they are integrated into the proposed picking tours.

3.6 Objective and Constraints

The problem consists in simultaneously determining for each period $t \in \mathscr{T}$: (1) the partition of each demand $d \in \mathscr{D}_t$ into boxes; (2) the batching of boxes into groups of up to f boxes, i.e., the determination of a picking tour in order to minimize the number of boxes that violate the minimal volume constraint and the distance covered by the pickers while satisfying the maximal volume and weight constraints, the walking policy imposed into the picking zone, and the trolley capacity.

Note that the objectives are prioritized. First, we minimize the number of boxes that violate the minimal volume constraint, and second, the walking distance. Moreover, even if the objective is to minimize the total distance walked by the pickers, the industrial partner wants to ensure the number of tours remains reasonable. Indeed, since there is not a unique starting location (a tour can start at any place along the conveyor belt), minimizing the distance could lead to generate a large set of small tours. This is not acceptable since starting a new tour requires some time for the picker, that can be viewed as a fixed cost. Hence, we impose a fixed distance penalty on all tours.

4 Algorithmic Approach

In this section, we describe the algorithm that has been developed to tackle the problem. Our algorithm includes two phases. The first phase splits each demand into different boxes (Sect. 4.1) while the second phase generates picking tours once the boxes have been determined (Sect. 4.2).

It can be noticed that the routing phase is clearly polynomial. Given a set of n locations to visit, because of the particular walking policy, the best route is determined by sorting the locations according to the walking order. This can be performed in $\mathcal{O}(n \log n)$. Hence the hardness of the problem is in the double packing: items of an order are packed into boxes, and boxes are grouped into tours.

4.1 Splitting Demands

Given a particular demand d, it may happen that the items cannot be contained into a single box or it may be advantageous to consider several boxes to increase the pickers' productivity. In these cases, the demand needs to be divided into several boxes able to contain it.

To perform this task, we apply the following algorithm inspired from the procedure *split* proposed by [1] and successfully used by [6] in the context of the vehicle routing problem. This procedure is based on the computation of shortest paths on acyclic graphs. The algorithm we propose, seeks for a distribution of a given demand in several boxes such that volume and weight constraints are respected while increasing productivity, i.e. yield to operational plans with the shortest possible walking distance.

In order to lighten the notation, the indices of demand d are removed for the products and their associated quantity. Hence, let us indicate by $\{p_1, p_2, \ldots, p_n\}$ the products required to satisfy demand d. The quantity associated with a product p_i is then denoted by q_i. It follows that demand d requires a total of $I = \sum_{i=1}^{n} q_i$ items.

Let us assume that each item $i \in \mathcal{I}$, $\mathcal{I} = \{1, \ldots, I = \sum_{i=1}^{n} q_i\}$ composing demand d has to be collected from the location l_i of the picking zone and let us suppose, without loss of generality, that items are ordered following the walking policy, namely such that $i, j \in \mathcal{I}, i < j$ implies $l_i \prec l_j$ (see definition in Sect. 3.2).

The total volume of demand d is $v_d = \sum_{i=1}^{n} q_i v(p_i)$, and the total weight of demand d is $w_d = \sum_{i=1}^{n} q_i w(p_i)$. If $v_d < v_{box}^{min}$ and $w_d \leq w_{box}^{max}$, all items of demand d are put into a single box that violates the minimal volume constraint. In the following, we only consider demand d such that $v_d \geq v_{box}^{min}$ or $w_d > w_{box}^{max}$. It means that the items of such demands may be split into several boxes.

We construct a graph $\mathcal{G} = (\bar{\mathcal{I}}, \mathcal{E})$ where $\bar{\mathcal{I}}$ is the node set while $\mathcal{E}$ is the arc set. Set $\bar{\mathcal{I}}$ contains $1 + I$ nodes, numbered $0, 1, \ldots, I = \sum_{i=1}^{n} q_i$. Node 0 is a dummy node, while node $i \in \bar{\mathcal{I}}, i > 0$ represents the i-th item in $\mathcal{I}$ forming demand d.

With each node $i \in \bar{\mathcal{I}}$, we associate the quantities v_i (resp. w_i) the volume (resp. the weight) of the item represented by node i.

Then, for each couple of nodes $i, j \in \bar{\mathcal{I}}$ we calculate three quantities:

- d_{ij} the walking distance needed to pick items $i + 1, \ldots, j$ from their locations $l_{i+1}, \ldots, l_j$:

$$d_{ij} = d_{0,l_{i+1}} + \sum_{h=i+2}^{j} d_{l_{h-1},l_h} + d_{l_j,0};$$

- $w_{ij} = \sum_{h=i+1}^{j} w_h$, i.e., the weight of items $i+1, \ldots, j$;
- $v_{ij} = \sum_{h=i+1}^{j} v_h$, i.e., the volume of items $i+1, \ldots, j$.

Based on these quantities, we can determine the arcs of graph $\mathscr{G}$ in such a way that each arc represents a feasible box. We first determine the arc set $\mathscr{E}_1$ defined as:

$$\mathscr{E}_1 = \{(i, j) | i < j, \ i, j \in \bar{\mathscr{I}}, v_{box}^{min} \leq v_{ij} \leq v_{box}^{max}, w_{ij} \leq w_{box}\}. \quad (1)$$

Second, in order to guarantee the graph connectivity, we consider the arc set $\mathscr{E}_2$ made of all arcs $(i-1, i)$, $i \in \bar{\mathscr{I}} \setminus \{0\}$ even if the associated box does not respect the constraint on the minimal volume. Thus,

$$\mathscr{E}_2 = \{(i-1, i) | i \in \bar{\mathscr{I}} \setminus \{0\}, v_{i-1,i} < v_{box}^{min}\}. \quad (2)$$

We finally define the arc set $\mathscr{E} = \mathscr{E}_1 \cup \mathscr{E}_2$.

The cost c_{ij} associated with each arc $(i, j) \in \mathscr{E}_1$ is given by the quantity $\alpha^{BOX} + d_{ij}$, where α^{BOX} is a parameter that penalizes the creation of new boxes. This parameter α^{BOX} permits to limit the number of boxes, hence to propose a reasonable number of tours. The cost c_{ij} associated with each arc $(i, j) \in \mathscr{E}_2$ is given by $\delta^{BOX} + d_{ij}$, where δ^{BOX} is a large-enough value.

To determine the distribution of the demand d into boxes, we compute the shortest path from the dummy source node 0 to node I with respect to costs c_{ij} . Each arc belonging to the shortest path represents a box. The shortest path provides the best split of products into boxes given the order of products and the cost structure.

The value of δ^{BOX} should be such that an arc in $\mathscr{E}_2$ is selected only if needed to guarantee the existence of a path between nodes 0 and I. Whenever the shortest path contains an arc in $\mathscr{E}_2$ the corresponding box does not respect all the constraints.

The algorithm is repeated for each demand $d \in \mathscr{D}$ to obtain its distribution into boxes. All the boxes that are created are inserted in set $\mathscr{B}$.

An example of the split procedure is given in Fig. 3a, b. Here the demand is composed of six different products noted a, b, c, d, e, f. They are located in the warehouse as illustrated in Fig. 3a. Values on the ordinate axis represent the distance from the beginning of the aisle to the product location, while values on the abscissa represent intra-aisle and inter-aisle distances. Following this disposition, we have that $a \prec b \prec c \prec d \prec e \prec f$. We suppose that only one item per product is required. For simplicity, let us suppose that all the products have a volume and a weight equal to 10 units, the minimal volume for a box equals 10, while the maximal weight and volume equal 30 units.

The graph that is built with respect to this demand is reported in Fig. 3b. A box that contains only product a is associated with a distance of $50 + \alpha^{BOX}$ units. The picker needs to get into the aisle and walk for 20 units to reach product a. He/she

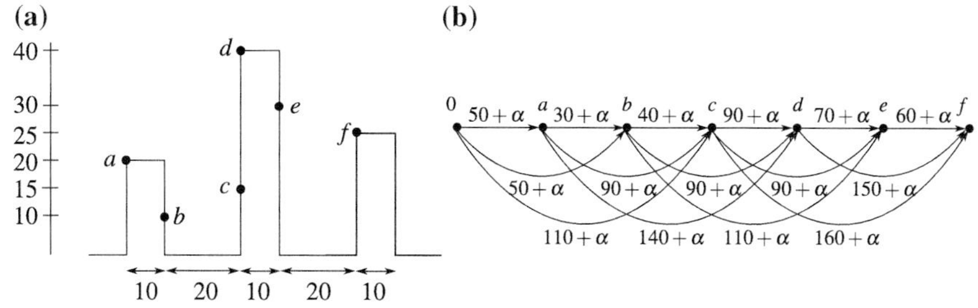

Fig. 3 Example of the split procedure. **a** Data for the example. **b** Graph for split procedure

then needs to cross and to get out from the aisle for an additional distance of 30 units. This is represented by arc $(0, a)$. A box that contains products b and c is associated with a distance of $90 + \alpha^{BOX}$, that is the distance to get into the first aisle, walk until the position in front of b, cross the aisle to reach b, get out from the aisle, and walk to the next aisle for a total of 50 units. Then, the picker needs to get into the second aisle and walk until c. He/she then crosses the aisle and gets out of it for an additional 40 units of distance. Thus the total distance to compose the box is $90 + \alpha^{BOX}$. This is represented by arc (a, c).

Note that if $\alpha^{BOX} \leq 50$ the shortest path that goes from node 0 to node f is made by arcs $(0, b)$, (b, e) and (e, f) for a total cost of $220 + 3\alpha^{BOX}$. The demand is then split in three boxes. One containing products a and b, another with products c, d and e, the last with only product f. On the other case, when $\alpha^{BOX} > 50$, the best solution is to create only 2 boxes. The shortest path in this case is made by arcs $(0, c)$ and (c, f) for a total cost of $270 + 2\alpha^{BOX}$. The solution in this case considers two boxes, one with products a, b and c, the other with products d, e and f.

4.2 Batching Boxes for Picker Trips

A picker pushes a trolley containing up to f boxes. We thus need to batch the boxes in order to obtain the trips that pickers have to perform. Let us suppose that a set $\mathcal{B} = \{1, \dots B\}$ of boxes have to be prepared. Let a_b^S be the starting aisle associated with box $b \in \mathcal{B}$, namely the aisle where is located the first item to pick to prepare box b (with respect to the walking policy). Let a_b^T be the ending aisle associated with box $b \in \mathcal{B}$: the aisle where is located the last item to pick to prepare box b. Moreover, let d_b be the walking distance to pick all items of box b. We suppose, without loss of generality, that boxes are numbered such that $b_1 < b_2, b_1, b_2 \in \mathcal{B}$ if and only if

$$\begin{aligned} a_{b_1}^S \neq a_{b_2}^S \text{ and } a_{b_1}^S \prec a_{b_2}^S \quad &\text{or} \quad a_{b_1}^S = a_{b_2}^S \text{ and } a_{b_1}^T \prec a_{b_2}^T \\ &\text{or} \quad a_{b_1}^S = a_{b_2}^S \text{ and } a_{b_1}^T = a_{b_2}^T \text{ and } d_{b_1} > d_{b_2}. \end{aligned}$$

To create batches of boxes of cardinality up to f, we adapt the split procedure described in Sect. 4.1. Similarly, a graph $\mathscr{G} = (\bar{\mathscr{B}}, \mathscr{E})$ is constructed. The node set $\bar{\mathscr{B}}$ contains $1 + B$ nodes, numbered from 0 to B. Node 0 is a dummy node, while node $i = 1, \ldots, B$ represents the i-th box in $\mathscr{B}$. The arc set $\mathscr{E}$ contains all the possible batches of boxes, namely, all batches with up to f boxes. Thus

$$\mathscr{E} = \{(i, j)|i, j \in \bar{\mathscr{B}}, j - i \leq f\}$$

A cost c_{ij} is associated with each arc $(i, j) \in \mathscr{E}$ and represents the total walking distance needed to collect all items in boxes $i + 1, i + 2, \ldots, j$, plus a fixed penalty α^{TOUR} to start a tour. Note that c_{ij} can be computed keeping in memory all the aisles that contain items in the boxes under consideration and the location of the furthest item in each aisle.

The shortest path that goes from the dummy node 0 to node B provide a partition of the box set $\mathscr{B}$ into batches of up to f boxes that minimizes the walking distance covered by the pickers.

4.3 Homogeneous Boxes

The boxes contain clothes that are exhibited in shops. When a box arrives at the shop, a salesperson carries it around the shop to prepare the displays. His/her job is made easier when boxes are *homogeneous*. Homogeneous means that a box contains clothes of a single class. Classes can be defined as needed, but an example of classes can be t-shirts, jeans, shirts, pulls, etc. For instance, a box containing 10 t-shirts is homogeneous, while a box containing one jeans and nine t-shirts is not.

Whenever homogeneity of the boxes is prior in the composition criteria, we propose to modify the split algorithm proposed in Sect. 4.1 to take this feature into account. The arc set defined by Eq. (1) does not change. However, we modify the cost associated with each arc by increasing it by a factor β whenever it considers a non-homogeneous box. Thus, the cost of an arc (i, j) such that items $i + 1, \ldots, j$ do not belong to the same class is $\alpha^{BOX} + \beta + d_{ij}$. When the value of β is increased, more homogeneous but also more numerous boxes are generated (especially when β and α^{BOX} are similar).

Moreover, if the minimal number of boxes required to contain the demand of a given customer, namely,

$$n_{box}^{min} = \max\left\{\left\lceil \frac{\sum_{i \in \mathscr{I}} v_i}{v_{box}^{max}} \right\rceil, \left\lceil \frac{\sum_{i \in \mathscr{I}} w_i}{w_{box}^{max}} \right\rceil\right\}$$

is greater than a certain parameter γ, the order on which items are considered to build the acyclic graph is modified and does not depend anymore on the walking policy. In particular, items are first ranked such that items of the same class follow each other, and inside each class, items are ranked according to the walking policy.

This increases the chance of having homogeneous boxes with the disadvantage of deteriorating the total walked distance.

5 Computational Results

Our industrial partner provided us with a set of 487 instances. These instances come from 17 working days between January and November 2017. Working days are divided into 11–16 periods. For each period, one instance is created for the demands to be prepared in blocks 1 and 2, and another instance is created for the demands to be prepared in blocks 3 and 4. For each instance, the set of demands corresponds to a number of items to be picked that varies between 45 and 18 209, with an average of 2 469.

We run the algorithms presented in Sects. 4.1–4.2 on the set of 487 instances to evaluate its efficiency. The parameters are set to keep a reasonable number of boxes and tours in the solution. Their value are: $\alpha^{BOX} = 30\,000$ and $\alpha^{TOUR} = 15\,000$. Each instance is solved on a Intel(R) Core(TM) i7-4600U CPU@2.10GHz, in less than one second. The overall results are reported in Table 1. From Table 1 we can notice that the number of boxes and the number of picking tours are similar in the solutions computed and in the solutions performed by HappyChic. However, the number of boxes with a volume lower than the minimal volume is drastically reduced in the proposed solution. Moreover, our approach drastically reduces the distance walked by pickers and, consequently increases their productivity. When we consider all the 487 instances individually, the proposed solution reduces the total distance in 462 instances. The deviation from the current industrial solution ranges from −90% to +8%, with an average of −21.85%. Hence the proposed solutions lead to significant improvements for the industrial partner.

This improvement can be explained by considering the current procedure of the industrial partner. HappyChic uses a greedy algorithm based on the following rule: given a demand, items are considered according to the walking policy. The first items are placed in a first box until the box is full. Then, a new box is opened and used

Table 1 Global comparison with the industrial solutions

	Industrial solution	Proposed solution	Deviation in percentage (%)
Nb. of boxes	47 201	47 354	+0.32
Nb. of picking tours	8 105	8 103	−0.02
Nb. box violate volume min	575	33	−94.26
Total walking distance (m)	1 298 958	984 479	−24.21
Nb. homogeneous boxes	12 463	13 978	+12.16
Pct. of homogeneous boxes	26.40%	29.52%	–

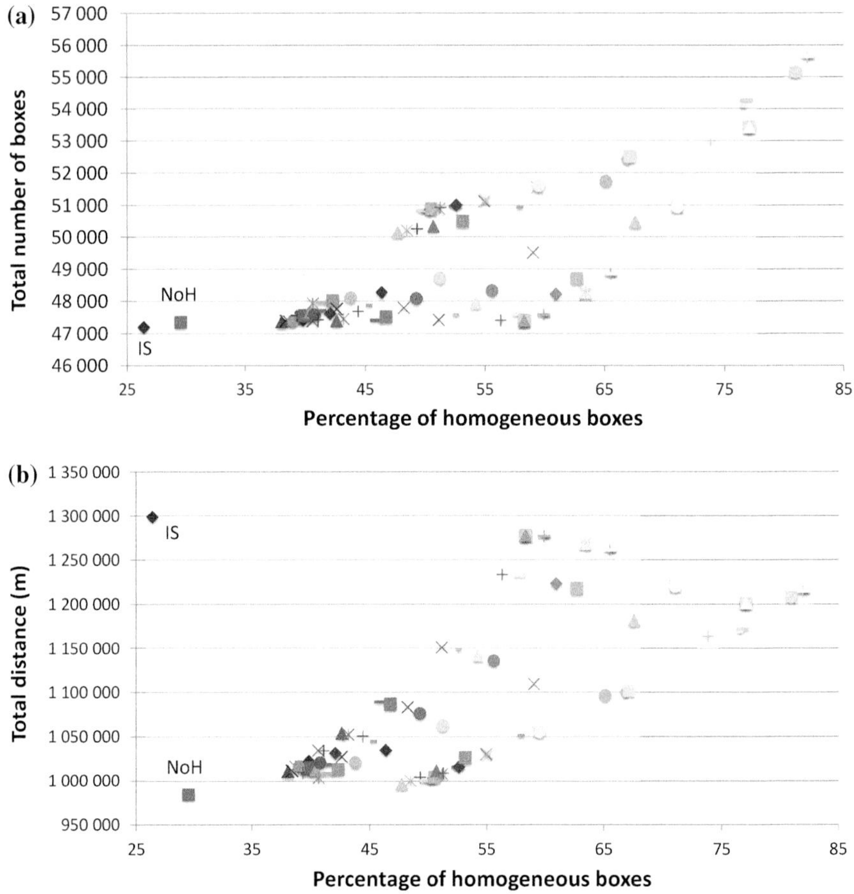

Fig. 4 Solutions taking into account homogeneity: each point is a solution. "*IS*" represents the current industrial solution and "*NoH*" represents the solution with taking the homogeneity into account. **a** Homogeneous boxes versus number of boxes. **b** Homogeneous boxes versus total distance

to collect other items. When it is full, the process is repeated. Hence it is likely that the initial boxes are filled up to capacity, but the last one may be almost empty. Moreover, boxes are built taking into account the walking policy but without taking into account distances among product locations. Our approach implicitly considers all feasible boxes taking into account the walking policy as well as distances leading to high improvements with respect to the current industrial solutions.

As explained in Sect. 4.3, the tasks of salespersons in shops are easier when boxes are *homogeneous*. In Fig. 4a, b we present results obtained by the modified split algorithm proposed in Sect. 4.3 when homogeneity is explicitly taken into account. The black dot labelled "*IS*" in the left part of the figures corresponds to the industrial solution which is the worst among the others in term of homogeneity. The solution obtained with the algorithm presented in Sect. 4.1 without taken homogeneity into account is represented by the gray square labelled "*NoH*" in the left part of the figures. This solution already improves the homogeneity of the boxes with respect to the industrial solution.

We have tested several values for parameters β and γ introduced in Sect. 4.3. β ranges from 5 000 to 40 000 with a step of 5 000, and γ ranges from 1 to 10 with a step of 1. Each combination of β and γ corresponds to a point in Fig. 4a, b.

The homogeneity of the boxes can be improved by applying the changes described in Sect. 4.3 at the cost of degrading the total walking distance and the number of boxes. It is possible to reach 35–45% of homogeneous boxes with a small increase in the distance and the number of boxes. This is achieved with $\beta \leq 20\ 000$ and $\gamma \geq 6$. Homogeneity can reach up to 82%, at an expensive cost in the number of boxes. However, with 82% of homogeneous boxes, the distance is still lower than the distance covered in the current industrial solution.

The logistic department of HappyChic is convinced by this research work and the results obtained. The warehouse management system is currently being updated in order to integrate the proposed algorithm.

6 Conclusions

In this paper, we presented an integrated order batching and picker routing problem based on an industrial case encountered in a warehouse of the company HappyChic. The specific features of the studied problem are: (1) orders have to be split into several boxes, (2) boxes are batched into trolleys, and (3) the picking zone has a specific configuration, so the optimal routing policy is easy to determine when the locations to visit are known. To deal with large size instances, we proposed a heuristic algorithm based on a double split procedure: the first one in order to create boxes, the second one to batch boxes into picking tours. The proposed algorithm can easily be adapted to take into account several classes of products with a homogeneity criterion when creating the boxes. The proposed results are very positive for the company: over a set of 487 instances, the total picking distance can be decreased by 24.21%.

Acknowledgements The authors warmly thank the HappyChic project team for this fruitful collaboration: Anne Grosset, Céline Brasseur, Emilie Delannoye, Elisabete Dierkens and Bruno Valleret. Thanks are also due to the referees for their valuable comments.

References

1. Beasley, J.: Route first-cluster second for vehicle routing methods. Omega **2**(4), 403–408 (1983)
2. de Koster, R., Le-Duc, T., Roodbergen, K.: Design and control of warehouse order picking: a literature review. Eur. J. Oper. Res. **182**(2), 481–501 (2007)
3. Ecommerce europe.: https://www.ecommerce-europe.eu/ (2017). Accessed 24 Oct 2017

4. Gademann, N., van de Velde, S.: Order batching to minimize total travel time in a parallel-aisle warehouse. IIE Trans. **37**(1), 63–75 (2005). https://doi.org/10.1080/07408170590516917
5. Marchet, G., Melacini, M., Perotti, S.: Investigating order picking system adoption: a case-study-based approach. Int J. Logist. Res. Appl. **18**(1), 82–98 (2015)
6. Prins, C.: A simple and effective evolutionary algorithm for the vehicle routing problem. Comput. Oper. Res. **31**(12), 1985–2002 (2004)
7. Scholz, A., Wäscher, G.: Order batching and picker routing in manual order picking systems: the benefits of integrated routing. Central Eur. J. Oper. Res. **25**(2), 491–520 (2017)
8. Valle, C.A., Beasley, J.E., da Cunha, A.S.: Optimally solving the joint order batching and picker routing problem. Eur. J. Oper. Res. **262**(3), 817–834 (2017). https://doi.org/10.1016/j.ejor.2017.03.069

Creation of Optimal Service Zones for the Delivery of Express Packages

Tiziano Parriani, Matteo Pozzi, Daniele Vigo and Frans Cruijssen

Abstract Service zones design is a crucial step in optimizing express delivery services which simplify the planning of everyday distribution routes while favoring the driver familiarity with the delivery zone. We describe a Decision Support System (DSS) for the definition of service zones based on an advanced multi-attribute clustering algorithm. The DSS has been used successfully by a major provider in The Netherlands to support the redesign of its distribution system for express parcel delivery.

Keywords Express delivery services · Logistics · Service zone design · Clustering

1 Introduction

Package delivery by express courier is a classical example of logistics problem in which many customers to be visited are scattered throughout a territory. In these problems, the territory is often subdivided into service zones of potential customers. In this way, specific resources (usually a vehicle and the associated driver) can be assigned to each service zone. The partition is often done with a strategical perspective and the zones remain stable for long periods of time. There are various motives for keeping the zones constant over time, such as:

T. Parriani · M. Pozzi
Optit srl, Imola, Bologna, Italy
e-mail: tiziano.parriani@optit.net

M. Pozzi
e-mail: matteo.pozzi@optit.net

D. Vigo (✉)
Optit srl and DEI, University of Bologna, Bologna, Italy
e-mail: daniele.vigo@unibo.it

F. Cruijssen
ArgusI, Breda, The Netherlands
e-mail: f.cruijssen@argusi.org

M. Dell'Amico et al. (eds.), *A View of Operations Research Applications in Italy, 2018*, AIRO Springer Series 2, https://doi.org/10.1007/978-3-030-25842-9_2

- **Familiarity with the customer**, i.e.,"Same person, at the same time". In a context where maximizing customer satisfaction has a key role in the service design, it is well known that the customer prefers that the deliveries be performed by a familiar person in an invariant time slot.
- **Familiarity with the service zone**. The delivery efficiency increases when the driver knows the service zone; such a knowledge can hardly be conveyed by a map. Very often it remains in the drivers mind, so only he or she knows all the tricks and shortcuts to best serve his or her zone.
- **Organizational and managerial simplicity**. From an organizational point of view, the stability of the service zones and appropriately assigned resources and the related sizing of the assigned resources makes the management of the delivery service easy.

With a relatively high frequency (typically daily) for each service zone, the points to be served are identified and one or more routes are defined. The routes will be performed by the vehicles and drivers assigned to the service zone. In practice, there are two main cases. In the first, the points in each service zone are regular customers which are served with a constant frequency: the routes themselves assume a strategical character because they are repeated virtually unchanged on specific days. In the second, which occurs frequently (for example, in parcel delivery services performed by express couriers), makes it necessary to re-plan the routes every day because the customers may be different from those served previously. As a result, the existence of service zones permanently assigned to drivers makes the routing easier, because drivers can individually decide the routes to be performed every day based on their deep knowledge of the territory. Thus, it is evident that familiarity with the service zone is an important factor for the operational effectiveness of the distribution system.

The design of the service zones is crucial, because it enables the subsequent definition of compact and stable routes. Such a planning approach groups together situations that are very different among them, from the road collection of urban waste to daily deliveries of packages. The collaboration between Optit Srl, an accredited spin-off of the University of Bologna, specializing in developing of Decision Support Systems (DSSs) for the management of complex systems, and ArgusI, a Dutch consulting company specialized in Supply Chain Optimization, led to the development of an innovative methodology, culminating in a complete DSS capable of overcoming the shortcomings of traditional approaches to service zones definition.

In many practical contexts, service zones are defined manually on the basis of geographical criteria or by referring to subdivisions designed by other operators for similar services—for example, subdivisions based on ZIP codes or city districts (see Fig. 1) or municipalities. Because this tends to disregard the peculiarities linked to the specific service to be performed, redesigning the service zones according to the specific needs of the service may lead to significantly better service quality. For an express courier, as we will see, defining service zones starting from ZIP codes can result in highly unbalanced workloads: some zones will have low vehicle saturation, thus needing to be aggregated with other zones for a specific day, while neighboring

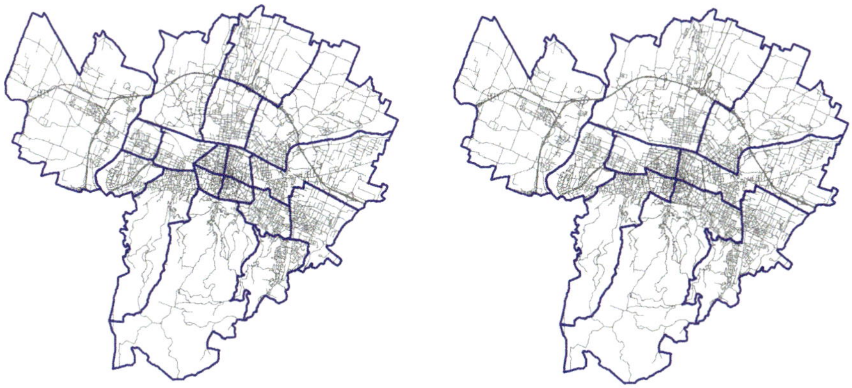

(a) districts (b) zones

Fig. 1 Example of a town's administrative districts (**a**), and potential service zones (**b**) based on aggregation of the districts

zones may instead require more than one vehicle to serve all their customers. In turn, this imbalance makes the process of defining routes more complex, requiring considerably more time and often producing sub-optimal decisions which result in inefficiencies in the service.

2 Optimization of Service Zones Definition

Some approaches to the optimal design of service zones for express parcel delivery have been presented in the literature (see, e.g., [1–3, 6, 8, 9]). However, they are generally so complex or time-consuming that the resulting DSSs lack the requisite flexibility: they are incapable of handling large-scale instances and the constantly changing requirements of real-world customers.

The method developed by Optit in collaboration with ArgusI subdivides a geographical macro-area served by an express courier into service zones according to certain requirements and parameters. In this case, historical data about the actual customer distribution drives the zone creation. We are given a set of potential customers to be served, each identified by a street address and the delivery characteristics. The zone is created by considering the travel times between the customers (points) and accounting for a set of additional constraints, such as: area and geometry of the service zones, maximum number of points permitted within each zone, estimated number of active points in a given day, estimated average time for delivery and estimated profit. The main objective of the service zone design phase is to minimize the number of zones. Since each zone is permanently associated with one or more vehicles, this objective is strongly related to the resources required to perform the service. However, other factors must be considered—such as the compactness of the zones, which is, in turn, strongly associated with the length and duration of the delivery routes. Thus, there are two, often conflicting, priorities: minimizing the number of zones while maximizing their compactness. The design of alternative solutions reconciles this conflict in different ways, allowing the exploration of alternative service scenarios. It is possible to define specific constraints as mentioned above, possibly taking into account the distance of the zone from the depot. Together, these priorities are aimed at sizing the service zones so that the daily routes utilize the available resources as fully as possible and the zones have balanced workloads

The developed heuristic algorithm, called OptiCluster, defines an optimized number of zones by iteratively solving a capacitated clustering problem as follows.

For a given number of clusters k, an initial relaxed solution is obtained with the well-known K-means algorithm (see, for example, [4]) to compute an initial set of zone centroids. Then, for each customer we create a "*wish-list*", which is an ordered list of pairs (cluster, centroid distance to customer). Each *wish-list* is sorted by non-decreasing order of the distance, and the difference between the second item and the first one defines the value of the so-called "regret function" for the customer. This value quantifies the distance penalty to be paid for serving that customer from the second-nearest centroid instead of the nearest. All the customers are then sorted by non-ascending order of regret functions, and they are assigned to the cluster on top of its *wish-list*. After each assignment, if other customers are no longer admitted to the cluster because of capacity limits, the relevant wish-lists and customer orders are updated, so that the customer with higher regret will always be assigned next. When all the customers have been assigned, the centroids are updated, the cluster is emptied, and the procedure is repeated with the new centroids. Like a K-means approach, the iterative process stops either when the position of the centroids does not change after all the customers have been assigned or when an iteration limit is reached. A local search step is then used to further reduce any cluster overlap and improve the zones workload balance. The final solution is further improved by using a ruin-and-recreate metaheuristic (see [7]) in which part of the solution is destroyed and rebuilt by using the regret-based construction method. This procedure can be repeated a number of times specified by an appropriate parameter, each time with different random centroids for the K-means algorithm. The result is an updated solution that maximizes the compactness of the given k clusters.

A valid upper bound for the number of clusters $\hat{k}$ is generated using an ad-hoc implementation of the agglomerative hierarchical clustering algorithm (see, e.g. [5]). Starting from a value k equal to $\hat{k}$, the procedure described above is executed to find near-optimal solution with k clusters. A cluster reduction is then performed by removing j clusters and destroying their neighbors. Clusters with fewer customers and lower values of their regret functions are favored for removal; the relative distance between clusters is also considered, in order to remove the sparsest clusters from the map. The new instance with $k - j$ clusters is solved, and the reduction is iterated until either a feasible solution is produced or cluster compactness is not worsened too much compared to the previous feasible solution.

OptiCluster is capable of solving the multi-dimensional clustering problems required for zone creation in a few minutes for instances with several thousands of

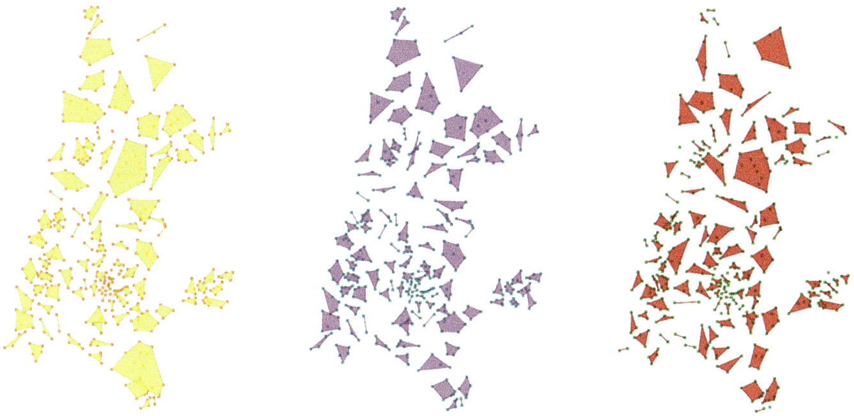

(a) 100 zones (b) 110 zones (c) 120 zones

Fig. 2 Convex hulls of the service zones for three alternative solutions obtained for the Amsterdam area by limiting the number of zones to 100, 110 and 120 for (**a**), (**b**) and (**c**), respectively. The points on the maps correspond to potential customers to be visited, while each polyline encloses a service zone

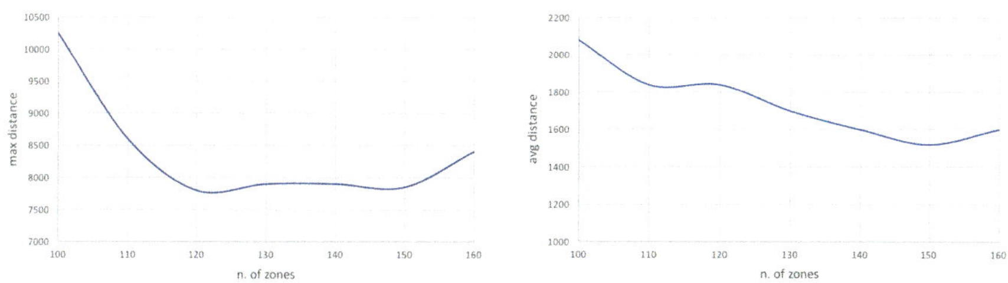

(a) max distance (b) average distance

Fig. 3 Trend line of the maximum (left) and average (right) distances between the points and the centroid of each service zone, obtained by varying the number of service zones in the solution

points. The service zones built by the algorithm inevitably require a manual adjustment before being actually put in place; however, extensive use of the algorithm in practical situations has shown that the amount of modification needed is very small. Moreover, a semi-automatic tool capable of quickly determining a set of service zones very close to those to be used in practice makes it possible for the planner to conduct extensive what-if analyses (considering and comparing different design rationales). For example, the impact of imposing a different number of target zones for a specific delivery region is illustrated in Figs. 2 and 3. Each solution was obtained by running the clustering step, which requires a couple of minutes for each run.

The Depot Optimization Tool (DOT), a complete DSS for service zone creation, incorporates Optit's OptiCluster multidimensional clustering algorithm within a user interface based on a Geographic Information System (GIS), (see Fig. 4). It was developed by ArgusI in 2013–14 and has been extensively used by the company in several

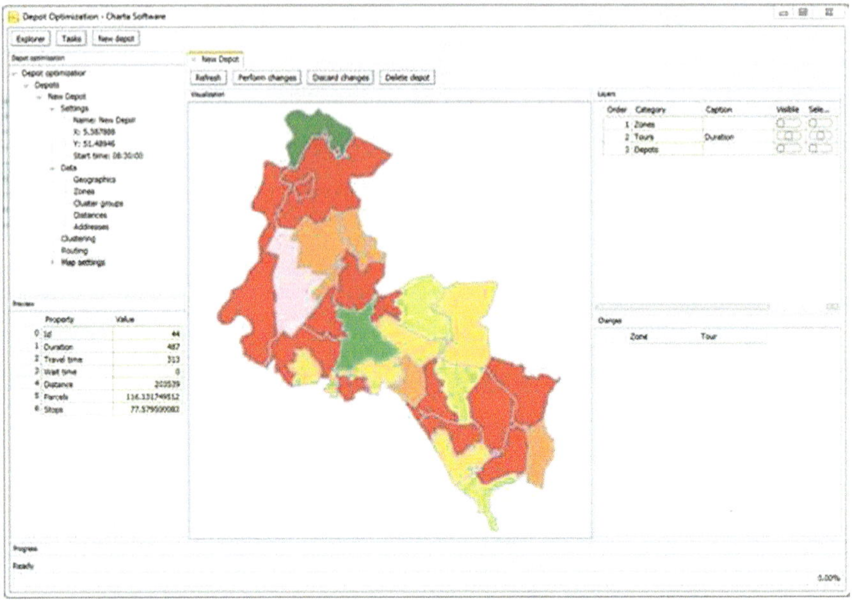

Fig. 4 The user interface of the DOT DSS by ArgusI, incorporating the Optit OptiCluster algorithm

projects since then. In addition, Optit also developed a specialized version of OptiCluster to create service zones in waste collection applications, which is currently used within the OptiRoute suite by several major Italian waste operators.

3 Case Study: Application in The Netherlands

From 2014 to 2016, ArgusI used DOT and OptiCluster to optimize the service zones for a major parcel-delivery provider in the Netherlands and Belgium (see Fig. 5). The design process for the ten depots of the Netherlands resulted in the creation of almost 9,000 service zones, starting from a sample of more than 120,000 addresses. Since then, the zones have been periodically revised to account for changes in demand patterns and service requirements, but the provider continues to use them daily.

To directly compare the benefits of the zones obtained by OptiCluster to zones created using traditional ZIP-based approaches, we consider, as an example, zone creation in the urban area of a large Dutch town. Figure 6 shows the typical uneven distribution of customers in the center; the density of delivery addresses often varies considerably across different areas of a town. Recall that the main objective of zone creation is to define relatively small zones, which in this case are associated with an estimated workload of approximately 30 min. Many such small zones are put together every day to create the actual delivery routes.

Fig. 5 Planned service zones for a major parcel delivery provider in the Netherlands

The service zones previously used by the courier were built using ZIP codes, with a different granularity of aggregation depending on the number of digits of the code used. We denote by PC4 and PC5 the zones obtained using four and five digits of the ZIP code, respectively. The 25 service zones defined by the PC4 aggregation level (see Fig. 7a) have a nice regular shape. Unfortunately, they are generally too large to be served by a single vehicle: their average workload is, in fact, close to 60 min. Furthermore, the uneven distribution of the demand across the different ZIP code zones means there is a large variation in the workload, as shown in Fig. 7b. Two zones have a workload of almost three hours and most of them have workload of about 60 min. Therefore, strong re-balancing actions need to be performed daily to build the actual delivery routes. When the PC5 level is used instead, the number of zones grows by one order of magnitude, reaching 274. The average workload (five minutes) is clearly much lower, but the solution is still very unbalanced, as shown by Fig. 7c, d.

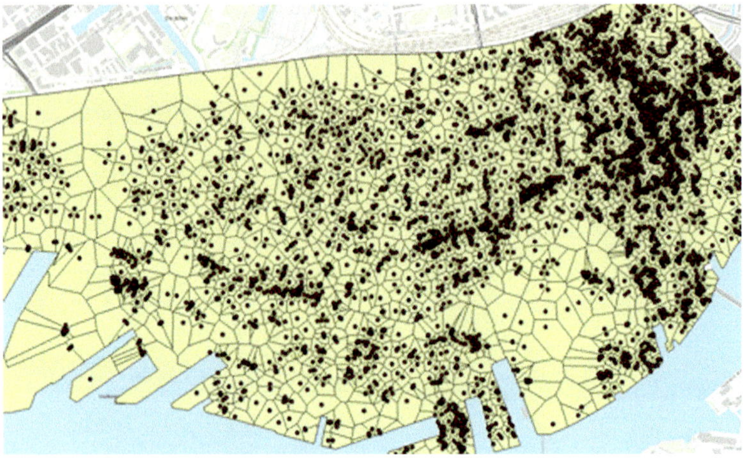

Fig. 6 The distribution of customers (black dots) in the center of a large Dutch town

(a) PC4 zones and their workload in minutes

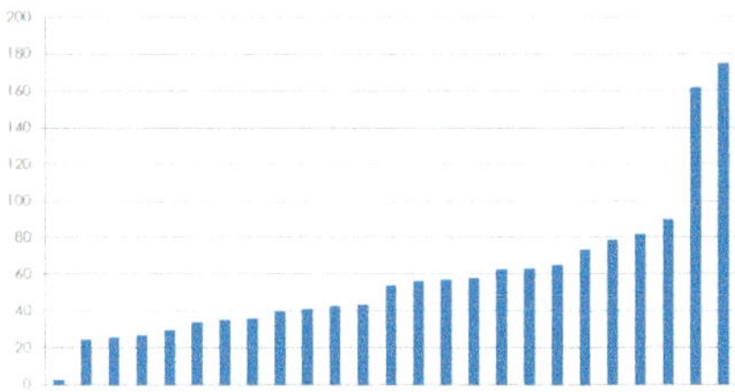

(b) Workload in minutes of PC4 zones

(c) PC5 zones and their workload in minutes

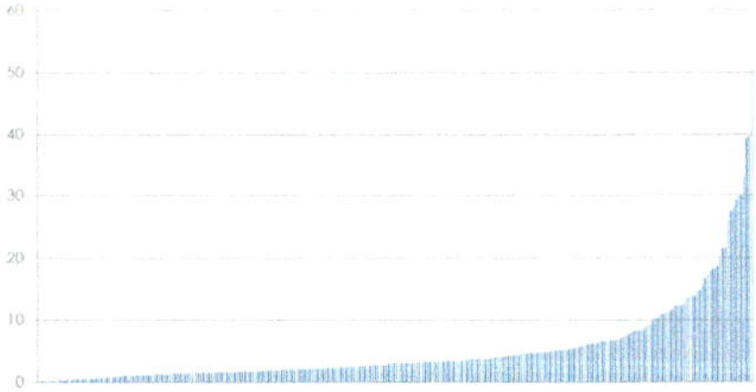

(d) Workload in minutes of PC5 zones

Fig. 7 Service zones defined by the PC4 and PC5 ZIP codes. In (**a**) and (**c**) the estimated workload in minutes of the zones is shown by different colors on a map, while in (**b**) and (**d**) the workload in minutes of the different zones is reported in ascending order

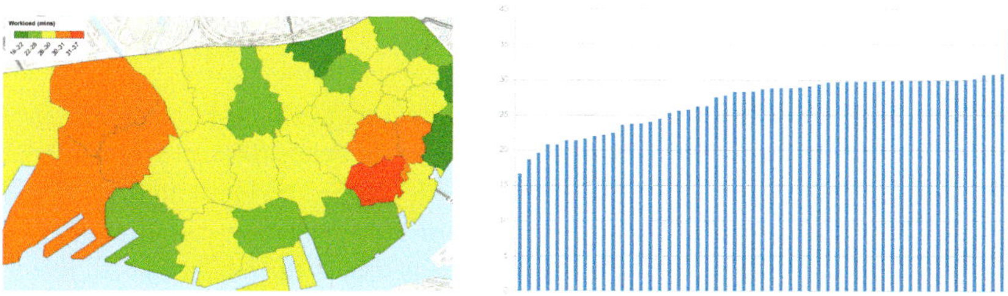

(a) OptiCluster zones and their workload in minutes

(b) Workload in minutes of OptiCluster zones

Fig. 8 Optimized service zones defined with OptiCluster algorithm within DOT. In (**a**) the estimated workload in minutes of the zones is shown by different colors, while in (**b**) the workload in minutes of the different zones is reported

Table 1 Comparison of optimized and ZIP-based zones

Zone type	# of service zones	Est. workload (min)		Coeff. of variation
		Average	Std. Dev.	
OptiCluster	54	27	4	0.15
PC4	25	58	39	0.67
PC5	274	5	7	1.40

The approach developed by Optit and ArgusI allows the user to set the preferred number of service zones together with a desired range for the estimated workload for each service zone, as well as several other constraints (such as a limited number of points). The final set of zones is obtained by using OptiCluster several times, varying the specific parameters until a satisfactory zone design is achieved. The automatic step is followed by a manual refinement step, in which the zones are slightly adjusted further to account for additional requirements which may be desirable. Figure 8 illustrates the solution obtained by OptiCluster prior to the final manual refinement. In this case, 54 zones have been built, and their average workload is about 27 min—very close to the target of 30 min. Furthermore, the variation of the zones workload is much smaller and can be easily adjusted with some manual updating of the solution. These results are summarized in Table 1, which reports the number of service zones and the average estimated service time for each zone along with the corresponding standard deviation and coefficient of variation (the ratio between the standard deviation and the average value of the workload). The table clearly shows the efficacy compared to traditional zones of those designed within a limited planning effort by using the DOT DSS. The use of DOT improved the effectiveness and reliability of the delivery service, greatly reduced the adjustments required to define the routes based on changeable daily demands and, last but not least, helped reduce the resources required to perform the deliveries.

4 Conclusions

We described the benefits of using clustering techniques based on state-of-the-art operations research to support the planning of service zones for express parcel delivery. The clustering algorithm OptiCluster was developed by Optit and incorporated by ArgusI into a complete GIS-based DSS for zone planning called DOT, which was used in several practical applications. We illustrated the advantages of the optimized zones achieved during a large project conducted for a major parcel-delivery provider in the Netherlands during 2014–16, by comparing them to ZIP-based ones. The OptiCluster algorithm has also been specialized by Optit for other specific applications such as waste collection in urban areas. Further, it is currently used in several practical planning projects.

References

1. Carlsson, J.: Dividing a territory among several vehicles. INFORMS J. Comput. **24**(4), 565–577 (2012)
2. Daganzo, C., Erera, A.: On planning and design of logistics systems for uncertain environments. In: New Trends in Distribution Logistics, pp. 3–21. Springer (1999)
3. Haugland, D., Ho, S., Laporte, G.: Designing delivery districts for the vehicle routing problem with stochastic demands. Eur. J. Oper. Res. **180**(3), 997–1010 (2007)
4. MacQueen, J.: Some methods for classification and analysis of multivariate observations. In: Proceedings of the Fifth Berkeley Symposium on Mathematical Statistics and Probability, Volume 1: Statistics, pp. 281–297. University of California Press, Berkeley, Calif (1967)
5. Murtagh, F.: A survey of recent advances in hierarchical clustering algorithms. Comput. J. **26**(4), 354–359 (1983). https://doi.org/10.1093/comjnl/26.4.354
6. Schneider, M., Stenger, A., Schwahn, F., Vigo, D.: Territory-based vehicle routing in the presence of time-window constraints. Transp. Sci. **49**(4), 732–751 (2014)
7. Schrimpf, G., Schneider, J., Stamm-Wilbrandt, H., Dueck, G.: Record breaking optimization results using the ruin and recreate principle. J. Comput. Phys. **159**(2), 139–171 (2000)
8. Wong, K., Beasley, J.: Vehicle routing using fixed delivery areas. Omega **12**(6), 591–600 (1984)
9. Wong, R.T.: Vehicle routing for small package delivery and pickup services. In: Golden, B., Raghavan, S., Wasil, E. (eds.) The Vehicle Routing Problem: Latest Advances and New Challenges, pp. 475–485. Springer, Boston, MA, USA (2008)

Solving a Three-Dimensional Bin-Packing Problem Arising in the Groupage Process: Application to the Port of Gioia Tauro

Luigi Di Puglia Pugliese, Francesca Guerriero and Roberto Calbi

Abstract In this paper we consider the consolidation operations in the groupage process. Containers, arriving from different origins carry objects with the same destination. We address the problem of inserting objects with the same destination into a new container. It is a three-dimensional bin-packing problem where several operational constraints are taken into account. We develop a simple but effective heuristic to solve the problem. This is based on the well-known next fit procedure where the concept of multi-layer is introduced. The defined heuristic allows to save empty space in the container dealing with a high impact on the real application. We present the main features of a decision support system that can help the logistic operators to implement a consolidation/deconsolidation hub.

Keywords Consolidation · Bin packing · Operative constraints · Heuristic

1 Introduction

In this paper, we address the consolidation problem in the groupage process. The term "groupage" is used to indicate a specific modality of goods and commodities's delivery, that consists in regrouping (consolidating) shipments from different suppliers, located in the same country, which have different destinations in another country. This shipment strategy (known in international sea freight transportation as "Less than container load") is the best solution for small shipments of generally up to 10 pallets, where the use of a whole container is not cost effective. Since the shipment does not fill a standard container, the container is loaded with cargo from

L. Di Puglia Pugliese · F. Guerriero (✉)
DIMEG, Universitá della Calabria, Rende (CS), Italy
e-mail: francesca.guerriero@unical.it

L. Di Puglia Pugliese
e-mail: luigi.dipugliapugliese@unical.it

R. Calbi
AutoTerminal Gioia Tauro S.p.A, Gioia Tauro, Italy
e-mail: calbi@blg-automobile.it

M. Dell'Amico et al. (eds.), *A View of Operations Research Applications in Italy, 2018*, AIRO Springer Series 2, https://doi.org/10.1007/978-3-030-25842-9_3

multiple consignees who share the container space. Upon arrival at the destination port, the goods are separated at a container freight station. This strategy allows a cost reduction for the customers, since they can order a limited quantity of items, resulting in a reduction of inventory cost and they can share the transportation costs. It is considered the most economical way to deliver small quantities of goods. The main aim of the consolidation problem is to group different objects in an optimal fashion, in order to allocate them orthogonally, without overlapping, into containers, by minimizing the number of used containers under several operational constraints. More specifically, the following restrictions are taken into account in the considered scenario:

- stacking constraints: it is assumed that some items are fragile, thus they cannot be stacked under other items [9];
- load balance constraints: items have to be loaded onto containers so that the stability of each containers is guaranteed [2];
- compatibility constraints: a determined couple of items cannot be placed in the same container [14];
- weight constraint: the sum of the weights of the loaded items must not exceed the weight limit imposed by the container [7];
- orientation constraints: only a single vertical orientation is allowed for each item [1].

It is assumed that

- all the items are available when the problem is solved;
- the cost to locate an item in a container is independent of the location.

From the mathematical point of view, the problem under study shares some features with well-known optimization problems, i.e., the three-dimensional multiple bin packing, the knapsack, and the container loading problems. Given the crucial role played by these problems in real settings, several contributions dealing with the development of efficient approaches for their solutions have been published in the scientific literature. We refer the reader to [3, 4] for recent surveys on the container loading problem and to [15] for mathematical models and solution approaches for different instances of the knapsack problem.

The scientific literature proposes several heuristics to solve the bin packing problem, i.e., local search procedures [8], genetic algorithm [6], and tabu search [10]. In [13] an approximation algorithm is presented. The reader is referred to [10, 12] for surveys on the three-dimensional bin packing problems.

In general, solution approaches for the three-dimensional packing problems are an extension of strategies for the bi-dimensional counterpart. In addition, the proposed metaheuristics in the literature make use of a constructive procedure in order to obtain initial solution. The most-known algorithms are the Next-Fit, First-Fit, and Best-Fit where the items are placed in non-increasing order of a predefined dimension. The three procedures differ from each other on the position in which the current item is located. It is worth observing that the aforementioned procedures have been designed for the bi-dimensional bin-packing problem. However, they can be tailored for the three-dimensional variant of the problem. In this case, the procedures generate a pile by inserting the next object over the previous inserted item. A new pile is generated when the current object, placed over the previous one, is outside the bin.

In [5], the authors define a two-phase approach. In particular, they consider a bin with an infinity length. In the first phase, the heuristic applies a constructive procedure obtaining a packing solution for the giant bin. Then, in the second phase, the number of bins needed to contain the items is determined.

In this work, we present a novel solution strategy for the consolidation problem, derived from the well-known Next-Fit heuristic developed for the classic bin packing problem (see, [11]).

The developed heuristic has been defined in strictly collaboration with the decision maker. It was easy to convince the decision maker about the validity of the methodology, since it is simple and very intuitive. In addition, the proposed algorithm results effective and able to solve real problems in a short computational time. For these reasons, we decided to propose an enhanced greedy approach instead of considering a more sophisticated methodology, available in the scientific literature.

The proposed algorithm is thus embedded in a decision support system, which can be used by logistic operators to execute the main operations arising in the management of a consolidation/deconsolidation hub.

The rest of the paper is organized as follows: in Sect. 2 we present the solution strategy developed for addressing the problem under study along with the description of the decision support system; in Sect. 3 some final conclusions are drawn.

2 A Constructive Heuristic: The Multi-layer Next-Fit Strategy

We develop a constructive heuristic to address the three-dimensional bin-packing problem. The main aim is to determine the packing of the objects in the containers (i.e., bins), inserting one object at a time into an appropriate position.

The proposed procedure can be viewed as a particularization of the Next-Fit Strategy (NFS, for short) [12] to the problem under examination, in which the concepts of multi-layer and sub-bins are introduced. The main idea is to consider each loaded item as a new (dummy) bin.

Before describing the developed procedure, it is useful to outline the main operations executed by the classic NFS, in which the items are initially sorted by nondecreasing values of a given dimension and packed in the corresponding sequence.

The NFS places the items in columns forming layers. The first layer coincides with the left hand side of the bin (see Fig. 1). When a layer has been completed, the objects will be positioned as if the bin's length is reduced by an amount equal to the maximum length, in the L dimension, of the objects packed on the last layer.

As can be seen from the diagram in Fig. 1, the length of the container is reduced and becomes L'. The gray space is the unused space. Consequently, even if there is

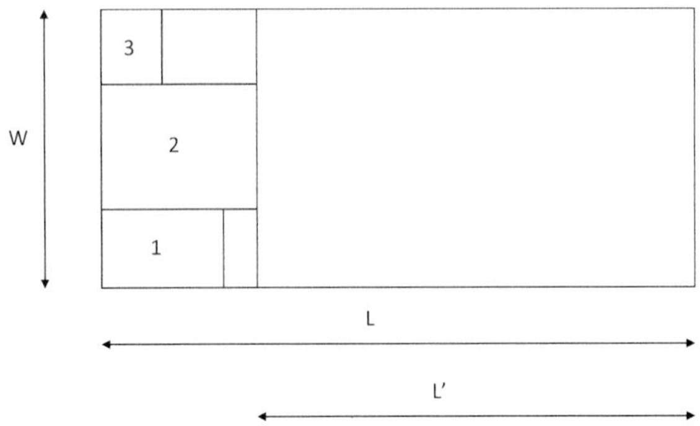

Fig. 1 The classic next fit strategy

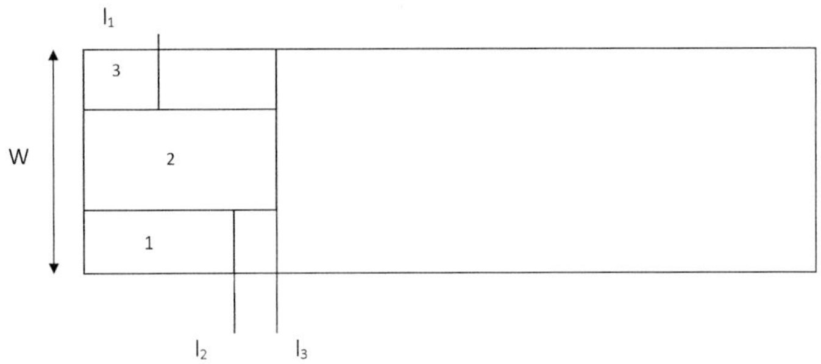

Fig. 2 The multi-layer next fit strategy

the possibility of placing an object in one of the unused spaces, this will be positioned starting from the left side of the container considering the new length L'.

The developed heuristic tries to minimize the unused space considering a multi-layer approach. A graphic representation of the multi-layer concept is provided in Fig. 2.

In particular, only if the object to be inserted cannot be positioned at the lower layers (i.e., l_1 and l_2), a new layer l_3 is open and the object is placed at l_3.

It is assumed that it is possible to rotate the items but only around the height, i.e., they can rotate of 90° in the plane $W - L$. This hypothesis is realistic because usually items are on a pallet so the orientation is given by the suppliers. Thus, the objects cannot be turned upside down.

The characteristics in terms of dimension (see Fig. 3) and weight are associated with each object o_k. In particular $l(o_k)$, $w(o_k)$, $h(o_k)$, $p(o_k)$ represent the length, width, height and weight of item o_k, respectively.

It is assumed that a set of heterogeneous bins is available. Four parameters (dimensions) are associated with each bin B_h, that is L_h which represents the length of the bin, W_h the width, H_h the height and P_h identifies the capacity. The coordinate

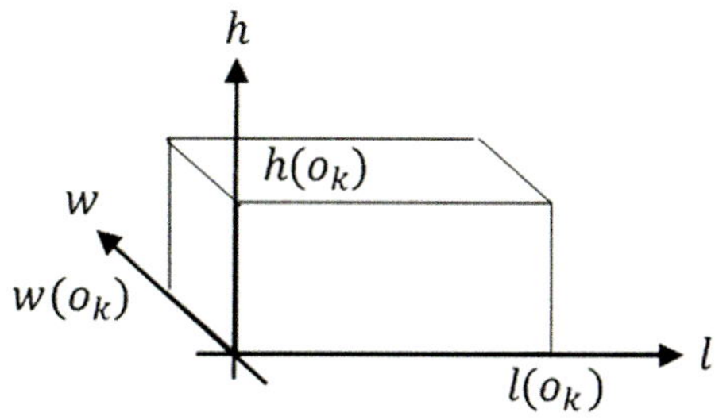

Fig. 3 Representation of the dimension of the object o_k

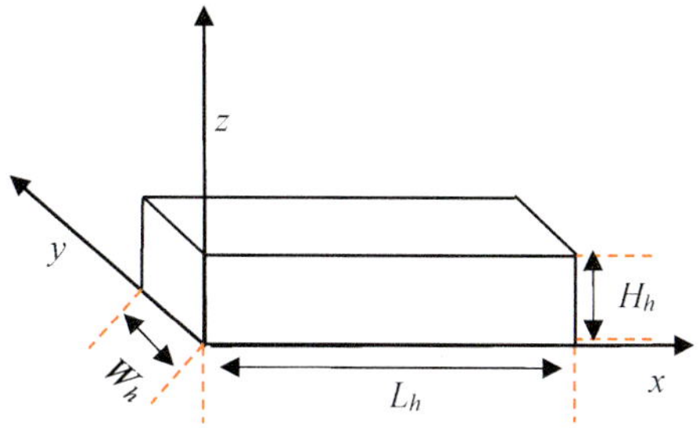

Fig. 4 Definition of the origin of the coordinate system in the bin B_h

system used for representing each bin B_h is depicted in Fig. 4. In particular, the origin of the system (x, y, z) is in the bottom left corner of B_h.

Let $Lay^h = \{l_m\}$; $m = 1, 2, \ldots,$ denote the set of layers of the bin B_h during the loading. The tuple $\left(\underline{Y}_m, \overline{Y}_m, X_m, h_m, I(l_m)\right)$ is associated with each layer l_m, where:

- the difference between $\underline{Y}_m$ and $\overline{Y}_m$ gives the maximum width of an object that can be placed in layer l_m;
- X_m defines the position of layer l_m with respect to the dimension L_h of the bin;
- h_m defines the position of the layer in terms of height, i.e., in reference to the H_h dimension of the bin;
- $I(l_m) = k$ identifies the object o_k that has generated the layer l_m.

It is assumed that the list of layers Lay^h is ordered in non-decreasing order of value of X_m.

Figure 5 provides the geometric representation that defines the parameters relating to the layers on the $W - L$ plane.

The following five variables are associated with each object o_k:

- x_k defines the position of the kth object with respect to the L axis of the bin;
- y_k defines the position of the kth object with respect to the W axis of the bin;
- z_k defines the position of the kth object with respect to the H axis of the bin;
- r_k define the orientation of the kth object. In particular, $r_k = 1$ if $l(o_k)$ is parallel to the dimension L of the bin, and $r_k = 0$ if $l(o_k)$ is parallel to the dimension W of the bin;
- b_K indicates the bin in which the object o_k has been placed.

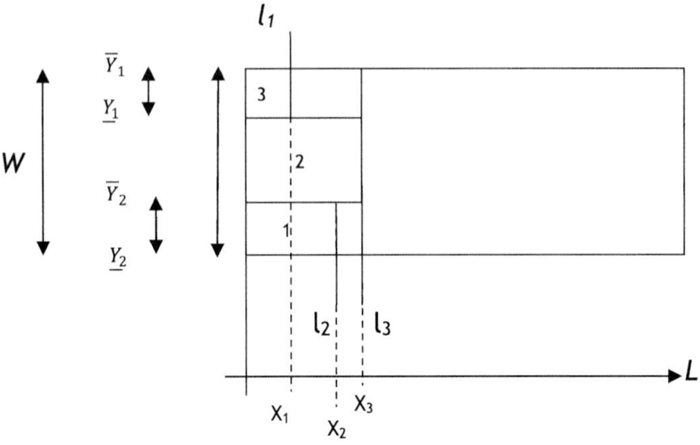

Fig. 5 Layer parameters. $I(l_1) = 3$, $I(l_2) = 1$ while $I(l_3) = 2$

We define with $Pcorr_h = \sum_{(o_k:b_k=h)} p(o_k)$ the weight of the load in the hth bin. To take into account the dimension H_h, it is assumed that each object o_k placed in a given layer l_m, is viewed as a sub-bin SB_h^q of the bin B_h. This bin will have dimensions $L_{q,h} = l(o_k)$, $W_{q,h} = w(o_k)$, $H_{q,h} = H_h - (h_m + h(o_k))$. The dimension $H_{q,h}$ represents the residual height of the bin B_h, considering that the object o_k is placed at a height $z_k = h_m$ and its height is equal to $h(o_k)$.

Let B be the set of the available bins. A list of layer Lay^h is associated with each bin $B_h \in B$. In the initialization phase, each list contains the layer $l_0 = (0, W_h, 0, 0, 0)$.

The proposed solution procedure considers the insertion of the objects into the bins, based on their placement in a list V_d, that contains objects with the same destination d and the same delivery date. The destination of the item i is defined as d_i. At least one list V_d is associated with each destination d. Therefore, the list with all the items i such that $d_i = d$ is indicated with V_d.

A matrix M is used to represent the incompatibility among the different types of goods. In particular, given two items i and j, if $M(i.t, j.t) = 1$ then the two items are incompatible.

Given an item i with $d_i = d$, i is inserted in V_d if and only if for each $j \in V_d$, $M(i.t, j.t) = 0$. If for some $j \in V_d$, $M(i.t, j.t) = 1$, then a new list V_d^1 is opened in which the item i is inserted.

In addition, the items are grouped based on the delivery date. More specifically, a list is created for each week (Monday to Friday) so that items with delivery date in week α are not included in the list of items with delivery date in week β. On the basis of the considerations reported above, it is evident that for each destination, different lists, containing compatible items, can be obtained. The items of each list are ordered for a non-decreasing value of the occupied area.

Moreover, because of the presence of fragile items, which cannot be stored in block stacks, a list $V_d^h(fragile)$ is associated with the list V_d^h. The list $V_d^h(fragile)$ will contain the fragile items with destination d and compatible with the items belonging to V_d^h.

The items of $V_d^h(fragile)$ will be placed on the items belonging to the V_d^h already processed (i.e., placed in the bins).

Let V_d^h be a list of items to be placed in the bins and let SB be the set of the sub-bins. We indicate with m^1 and m^2 the best layers where placed an object o_k when r_k is equal to 0 and 1, respectively. The fundamental steps performed by the proposed heuristic method, can be summarized in Algorithm 1.

From the description of Algorithm 1, it is evident that once a bin to be filled is selected, the algorithm minimizes the unused space, through the management of the multi-layer. Moreover, the introduction of the sub-bins ensures that an object can be inserted in a bin B_h already considered in the previous iterations, but not yet completely filled. In particular, Step 1 manages the order in which the bins must be "open", while Step 1.1 manages unused spaces in already "open" containers. If the object selected in Step 2 does not enter any sub-bin, i.e., it is not possible to insert the object in an already considered bin, then a new one is opened. The main objective of the proposed heuristic is to minimize the total number of used bins. In this respect, the best choice is to select the bins in non-decreasing order of their volume.

2.1 A Decision Support System

The solution approach presented in Sect. 2 is the core of a decision support system (DSS, for short) that can be used by the logistic operator, to solve the consolidation problem.

The system is characterized by three modules:

- a user interface to facilitate the input of the data and the analysis of the generated solutions;
- the optimization module, which implements the Multi-layer Next Fit heuristic;
- a database for the storage and the retrieval of the data.

In order to obtain the containers configuration, first of all, the user has to upload the items to be shipped, by using the order entry modality. Figures 6 and 7 show a screen-shot of the order entry modality. For readability reason, we divided the screen-shot in two different figures. The gray rectangles hide sensible data, i.e., Client, Shipper, Consignee, Degroupage Destination, and Destination. Figures 6 and 7 depict one of the instance, provided by a logistic operator, operating at the Gioia Tauro harbor, the largest port in Italy for container throughput, located north of the city of Reggio Calabria, between the municipalities of Gioia Tauro and San Ferdinando, in the Calabria region (South of Italy).

The items, loaded in the order entry modality, are consolidated by clicking on the button "Create Bins". The heuristic is launched and the result is displayed to the user as depicted in Fig. 8.

Algorithm 1 Multi-layer Next Fit Procedure

Step.0 (Initiliazation)
Determine the set V_d^h
Set $b_k = x_k = y_k = z_k = 0; r_k = 1; \forall o_k \in V_d^h$
Step.1 (Bin Selection)
Select a bin B_h from the list B;
Consider B_h as a sub-bin B_h^0 and add it to SB
Step.1.1 (Sub-Bin Selection)
Select a sub-bin SB_h^q from the set SB
Step.2 (Object Selection)
if $V_d^h = \emptyset$ **then**
  return
else
  Extract the object o_k from the top of the list V_d^h
end if
Step.3 (Object Insertion)
Set $l_{m^1} = l_{m^2} = NULL$
Determine
$m^1 = arg\ min_{l_m \in Lay^{q,h})}\{X_m : \overline{Y}_m - \underline{Y}_m \geq r_k w(o_k) + (1 - r_k)l(o_k) \wedge X_h + r_k l(o_k) + (1 - r_k)w(o_k) \leq L_{q,h} \wedge z_k \leq H_{q,h} \wedge r_k = 0\}$;
Determine
$m^2 = arg\ min_{l_m \in Lay^{q,h})}\{X_m : \overline{Y}_m - \underline{Y}_m \geq r_k w(o_k) + (1 - r_k)l(o_k) \wedge X_h + r_k l(o_k) + (1 - r_k)w(o_k) \leq L_{q,h} \wedge z_k \leq H_{q,h} \wedge r_k = 1\}$;
Set $\bar{l}_m = min\{l_{m^1}, l_{m^2}\}$
if $\bar{l}_m \neq NULL \wedge Pcorr_h + p(o_k) \leq P_h$ **then**
  if $\bar{l}_m = l_{m^1}$ **then** set $r_k = 0$
  else set $r_k = 1$.
  end if
  Set $x_k = X_m$; $y_k = \underline{Y}_m$; $z_k = h_m$; $b_k = h$; $Pcorr_h = Pcorr_h + p(o_k)$;
  Create a new layer $l_n = (0, 0, x_k + r_k l(o_k) + (1 - r_k)w(o_k), z_k, k)$ and insert l_n into $Lay^{q,h}$;
  Create a new sub-bin $SB_h^{\bar{q}}$ with $L_{\bar{q},h} = l(o_k)$, $W_{\bar{q},h} = w(o_k)$; $H_{\bar{q},h} = H_h - (h_m + o_k(h))$;
  $Lay^{\bar{q},h} = \{l_0 = (0, W_{SB_h}, x_k, h_m + h(o_k), 0)\}$ and insert $SB_h^{\bar{q}}$ into SB;
  if $n = |Lay^{q,h}|$ **then** Set $\underline{Y}_m = 0$ and $\overline{Y}_m = W_h$
  if $n \neq |Lay^{q,h}|$ **then** Set
  $\underline{Y}_m = max_{o_h : h = I(l_v) \wedge v > n \wedge y_k > y_h}\{y_h + r_h w(o_h) + (1 - r_h)l(o_h)\}$ and
  $\bar{Y}_m = min_{o_h : h = I(l_v) \wedge v > n \wedge y_k < y_h}\{y_h\}$
  for $u < n$ **do**
    if $\underline{Y}_u < y_k \wedge \overline{Y}_u = W_h$ **then** Set $\overline{Y}_u = y_k$.
    if $\underline{Y}_u > y_k \wedge \underline{Y}_u < y_k + r_k w(o_k) + (1 - r_k)l(o_k)$ **then** Set $\underline{Y}_u = y_k + r_k w(o_k) + (1 - r_k)l(o_k)$
    if $\underline{Y}_u > \overline{Y}_u$ **then** Eliminate the layer l_u
  end for
  Go to Step. 2.
else if $\bar{l}_m = NULL \vee Pcorr_h + p(o_k) > P_h$ **then**
  If all the sub-bins belonging to SB have been processed **then** execute Step. 1
  else execute Step. 1.1.
  end if
end if

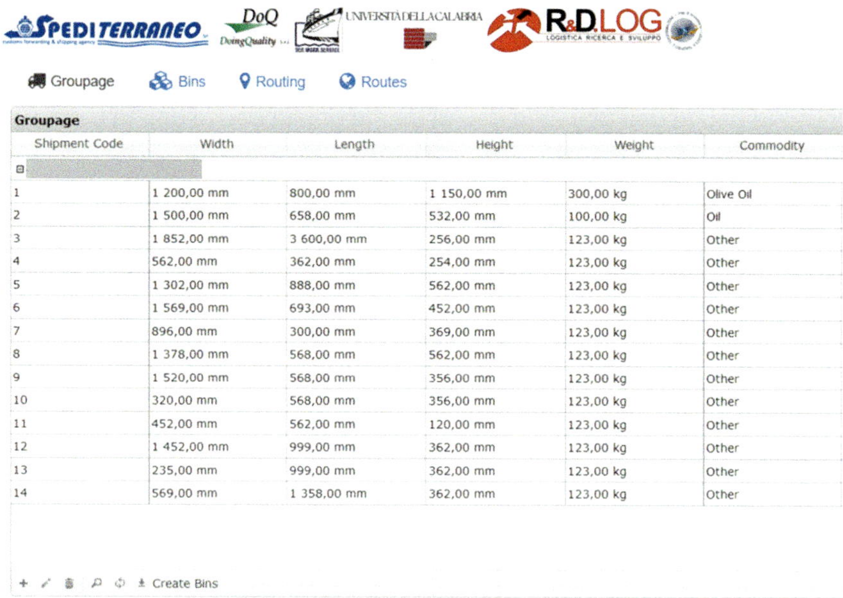

Groupage

Shipment Code	Width	Length	Height	Weight	Commodity
1	1 200,00 mm	800,00 mm	1 150,00 mm	300,00 kg	Olive Oil
2	1 500,00 mm	658,00 mm	532,00 mm	100,00 kg	Oil
3	1 852,00 mm	3 600,00 mm	256,00 mm	123,00 kg	Other
4	562,00 mm	362,00 mm	254,00 mm	123,00 kg	Other
5	1 302,00 mm	888,00 mm	562,00 mm	123,00 kg	Other
6	1 569,00 mm	693,00 mm	452,00 mm	123,00 kg	Other
7	896,00 mm	300,00 mm	369,00 mm	123,00 kg	Other
8	1 378,00 mm	568,00 mm	562,00 mm	123,00 kg	Other
9	1 520,00 mm	568,00 mm	356,00 mm	123,00 kg	Other
10	320,00 mm	568,00 mm	356,00 mm	123,00 kg	Other
11	452,00 mm	562,00 mm	120,00 mm	123,00 kg	Other
12	1 452,00 mm	999,00 mm	362,00 mm	123,00 kg	Other
13	235,00 mm	999,00 mm	362,00 mm	123,00 kg	Other
14	569,00 mm	1 358,00 mm	362,00 mm	123,00 kg	Other

Fig. 6 The DSS's order entry modality (first part)

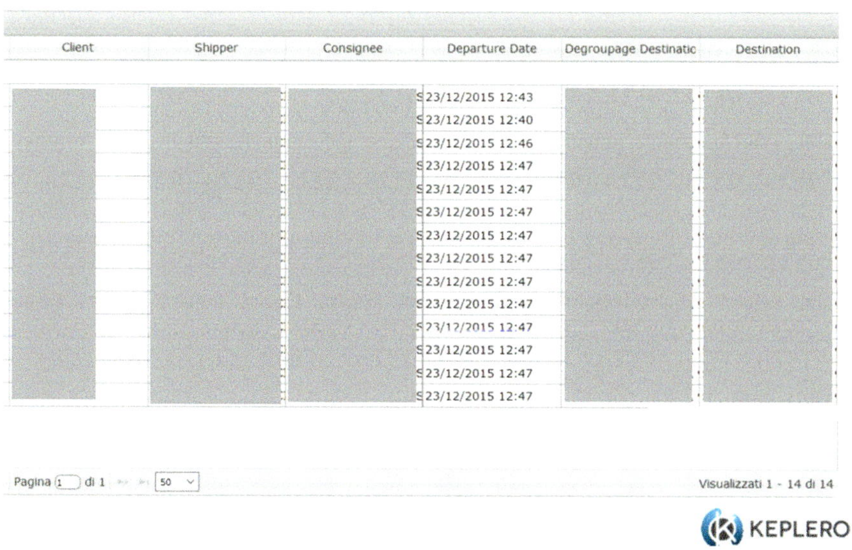

Client	Shipper	Consignee	Departure Date	Degroupage Destinatic	Destination
			23/12/2015 12:43		
			23/12/2015 12:40		
			23/12/2015 12:46		
			23/12/2015 12:47		
			23/12/2015 12:47		
			23/12/2015 12:47		
			23/12/2015 12:47		
			23/12/2015 12:47		
			23/12/2015 12:47		
			23/12/2015 12:47		
			23/12/2015 12:47		
			23/12/2015 12:47		
			23/12/2015 12:47		
			23/12/2015 12:47		

Fig. 7 The DSS's order entry modality (second part)

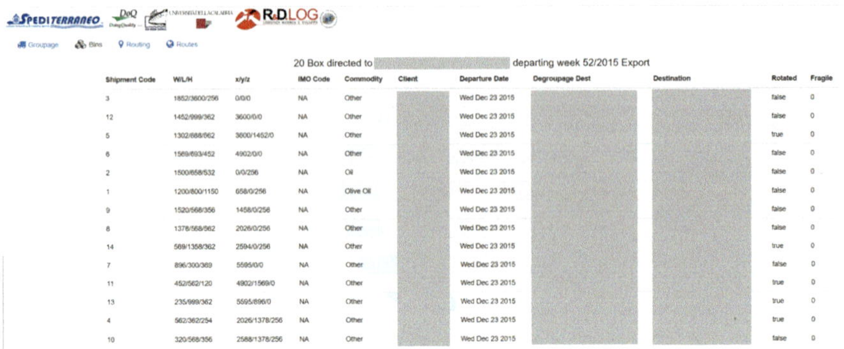

20 Box directed to departing week 52/2015 Export

Shipment Code	W/L/H	x/y/z	IMO Code	Commodity	Client	Departure Date	Degroupage Dest	Destination	Rotated	Fragile
3	1852/3600/256	0/0/0	NA	Other		Wed Dec 23 2015			false	0
12	1452/999/362	3600/0/0	NA	Other		Wed Dec 23 2015			false	0
5	1302/688/562	3600/1452/0	NA	Other		Wed Dec 23 2015			true	0
6	1569/693/452	4902/0/0	NA	Other		Wed Dec 23 2015			false	0
2	1500/658/532	0/0/256	NA	Oil		Wed Dec 23 2015			false	0
1	1200/800/1150	658/0/256	NA	Olive Oil		Wed Dec 23 2015			false	0
9	1520/568/356	1458/0/256	NA	Other		Wed Dec 23 2015			false	0
8	1378/568/562	2026/0/256	NA	Other		Wed Dec 23 2015			false	0
14	569/1358/362	2594/0/256	NA	Other		Wed Dec 23 2015			true	0
7	896/300/369	5595/0/0	NA	Other		Wed Dec 23 2015			false	0
11	452/562/120	4902/1569/0	NA	Other		Wed Dec 23 2015			true	0
13	235/999/362	5595/896/0	NA	Other		Wed Dec 23 2015			true	0
4	562/362/254	2026/1378/256	NA	Other		Wed Dec 23 2015			true	0
10	320/568/356	2588/1378/256	NA	Other		Wed Dec 23 2015			false	0

Fig. 8 The container configuration for the orders of Fig. 6

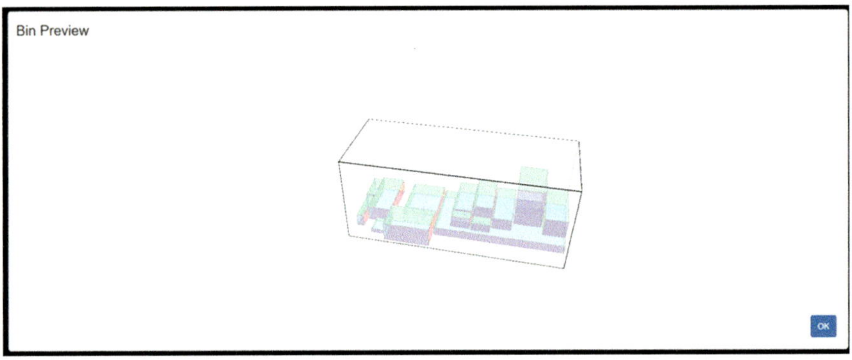

Fig. 9 3D visualization of the items placed in the container

The DSS gives the possibility of visualizing the item packed in the containers. The result, obtained from the data of Figs. 6 and 7, is shown in Fig. 9.

The proposed DSS has been used to solve some real consolidation problems, by a small Italian Forwarding Company and Co-Loader, operating at the Gioia Tauro port.

The results obtained for 2017 are very encouraging. Considering the time required to prepare the loading plan and to perform the loading operation as metric, the use of the DSS has allowed an average reduction of 1.5 h per 20 ft container to be achieved, passing from 5.0 to 3.5 h.

Considering the hourly cost of two workers, the depreciation of the forklift, and the energy cost, 1.5 h for preparing and performing the loading operation corresponds to an average cost of 90 euros per container. Thus the use of the DSS determined a cost saving of 90 euros per container, on average.

3 Conclusions

In this paper, we have addressed the consolidation problem, arising in the groupage process. The problem under study has been formulated and solved as a particular instance of the three-dimensional bin packing problem, where some operational constraints, dealing with real logistic processes, have been introduced.

A generalization of the Next-Fit heuristic, proposed for the classic two-dimensional bin packing problem, has been developed. The main idea is to introduce the concept of multi-layer, with the aim of minimizing the unused space in each container. In addition, the concept of sub-bins has been introduced and defined, in order to optimize the occupied space in the height dimension.

The algorithm has been embedded in a decision support system and it has been used to solve some real problems, at the port of Gioia Tauro.

Acknowledgements The authors are grateful to project Groupage (funded by the European Union and Calabria Region—POR FESR Calabria 2007/2013 Asse I Ricerca Scientifica, Innovazione Tecnologica e Societá dell'Informazione) for financial support of this research.

References

1. Amossen, R.R., Pisinger, D.: Multi-dimensional bin packing problems with guillotine constraints. Comput. Oper. Res. **37**(11), 1999–2006 (2010)
2. Bortfeldt, A., Gehring, H.: A hybrid genetic algorithm for the container loading problem. Eur. J. Oper. Res. **131**(1), 143–161 (2001)
3. Bortfeldt, A., Wäscher, G.: Constraints in container loading a state-of-the-art review. Eur. J. Oper. Res. **229**(1), 1–20 (2013)
4. Brandt, F., Nickel, S.: The air cargo load planning problem - a consolidated problem definition and literature review on related problems. Eur. J. Oper. Res. (2018)
5. Chung, F.K.R., Garey, M.R., Johnson, D.S.: On packing two-dimensional bins. SIAM J. Algebr. Discret. Methods **3**, 66–76 (1982)
6. Corcoran, A.L., Wainright, R.L.: A genetic algorithm for packing in three dimensions. In: Proceedings of the 1992 ACM/SIGAPP Symposium on Applied Computing: Technological Challenges of the 1990's, pp. 1021–1030 (1992)
7. Egeblada, J., Garavelli, C., Lisi, S., Pisinger, D.: Heuristics for container loading of furniture. Eur. J. Oper. Res. **200**(3), 881–892 (2010)
8. Faroe, O., Pisinger, D., Zachariasen, M.: Guided local search for the three-dimensional bin packing problem. INFORMS J. Comput. **15**(3), 267–283 (2003)
9. Junqueira, L., Morabito, R., Sato Yamashita, D.: Three-dimensional container loading models with cargo stability and load bearing constraints. Comput. Oper. Res. **39**, 74–85 (2012)
10. Lodi, A., Martello, S., Vigo, D.: Heuristic algorithms for the three-dimensional bin packing problem. Eur. J. Oper. Res. **141**(2), 410–420 (2002)
11. Lodi, A., Martello, S., Vigo, D.: Recent advances on two-dimensional bin packing problems. Discret. Appl. Math. **123**(1), 379–396 (2002)
12. Martello, S., Pisinger, D., Vigo, D.: The three-dimensional bin packing problem. Oper. Res. **48**(2) (2000)
13. Miyazawa, F.K., Wakabaysashi, Y.: An algorithm for the three dimensional bin packing problem with asymptotic performance analysis. Algorithmica **18**, 122–144 (1996)

14. Muritiba, A.E.F., Iori, M., Malaguti, E., Toth, P.: Algorithms for the bin packing problem with conflicts. INFORMS J. Comput. **22**(3), 401–415 (2010)
15. Wilbaut, C., Hanafi, S., Salhi, S.: A survey of effective heuristics and their application to a variety of knapsack problems. IMA J. Manag. Math. **19**(3), 227–244 (2008)

A New Software System for Optimizing the Operations at a Container Terminal

Tiziano Bacci, Stefano Conte, Domenico Matera, Sara Mattia and Paolo Ventura

Abstract This paper describes the procedures implemented in the software system developed for the CONTRAST project to optimize the logistic operations at a container terminal. In particular, we consider the problems of minimizing the number of reshuffle operations and designing the routes of the vehicles inside the yard. Minimizing the number of reshuffle operations required to empty a container yard is addressed in the literature as the Block Relocation Problem and it is known to be NP-hard. Here we implemented two heuristic procedures that provide feasible solutions to the problem when new containers enter the yard or when some container must be reallocated for any reason.

Keywords Container relocation · Container yard · Container terminal

1 Introduction

The logistic application described in the paper was studied within the POR-FESR Lazio project "CONtainer TRAnshipment STation (CONTRAST)". The aim of the project was the development of an optimization based logistic software system to handle logistic operations at container terminals. The principal investigator of the project was AreSoft S.r.l., whose expertise in the field comprised the development

---

T. Bacci · S. Mattia · P. Ventura (✉)
Istituto di Analisi ed Informatica "A. Ruberti" del CNR, Via dei Taurini 19, Rome, Italy
e-mail: paolo.ventura@iasi.cnr.it

T. Bacci
e-mail: tiziano.bacci@iasi.cnr.it

S. Mattia
e-mail: sara.mattia@iasi.cnr.it

S. Conte · D. Matera
AreSoft S.r.l., Viale Massaia 31, Rome, Italy
e-mail: stefano.conte@aresoft.it

D. Matera
e-mail: domenico.matera@digitalengineering.it

M. Dell'Amico et al. (eds.), *A View of Operations Research Applications in Italy, 2018*, AIRO Springer Series 2, https://doi.org/10.1007/978-3-030-25842-9_4

of a management software for one of the largest container terminal of Lazio, the one of Pomezia. The scientific partners of the projects where the Istituto di Analisi dei Sistemi ed Informatica (IASI) "A. Ruberti" of CNR, the University "La Sapienza" of Rome, and the University of Genoa. In particular, IASI was responsible of designing and implementing the optimization algorithms defined as callable functions of the software. The research group at University of Genoa analyzed the historical data of the container terminal of Pomezia (provided by AreSoft s.r.l.) and constructed a dataset of instances that was used for the computational experiments. Then, the research group at University "La Sapienza", took care of implementing the web services that interface the Microsoft SQL Server designed by AreSoft s.r.l. and the optimization algorithms mentioned above.

Container yards (or terminals) are complex infrastructures whose importance is increasing over time, due to the large number of goods requiring intermodal transportation (see [2]). Depending on the terminal (see Fig. 1 for a picture), it may allow to temporarily store containers and interchange goods between ships, trucks and trains. The logistic operators are required to handle the containers trying to minimize times and costs. Containers are moved by rail mounted gantry cranes and simple cranes, whose type may vary, depending on the yard and on the operations that are performed (see Fig. 2). The yard is usually divided into blocks, each of them possibly reserved to a specific type of containers. Each block is, in turn, divided into bays, that contain a set of stacks where the containers are allocated in tiers, which are filled according to a first-in first-out policy. When a container has to be retrieved from its current location, either to exit the yard or to be inspected by officers, all the containers that are allocated above it (*blocking containers*), must be relocated to another stack (*reshuffled*). Indeed, one of the main issues in container yards are the reshuffle operations of the blocking containers. A huge amount of literature concerning storage yard operations

Fig. 1 A container terminal

Fig. 2 A crane moving a container

in container terminals has been proposed in the last two decades, with an always increasing use of several optimization techniques. Surveys on such problems can be found in [7, 14]. In the reference yard considered in the project (the Pomezia terminal), the containers arrive by trains and exit by trucks, without ships involved. Each container has a given *type* label and an (expected, as in most real cases) exit time.

The terminal has about 300 stacks with *height* (in term of available tiers) 3 or 4 and each stack can allocate only containers of a certain type (among 6 total types of containers). Containers are allocated in the stacks by means of simple cranes (cranes, for short) and retrieved from the yard by cranes or trucks. Three main operations can be identified: (i) reallocating the containers that have to be reshuffled; (ii) locating the containers at their arrival in the yard; (iii) designing the routes for the vehicles (cranes and trucks) inside the yard. Each of these operations is associated with a function implemented in the CONTRAST software, namely CF1, CF2, and CF3. In particular, function CF1 is called every time a container has to be retrieved from its current position (because it has to leave the yard or for an officers inspection). Then, the corresponding blocking containers must be reshuffled and, according to the value returned by CF1, assigned to the other stacks of the terminal. It is important to remark that policy of Pomezia's terminal did not allow to reshuffle containers that are not blocking for the next one to be retrieved. Function CF2 is called when a new train of containers arrives at the yard. For each of such containers, CF2 returns the stack where it has to be allocated. In both functions CF1 and CF2, the assignment of containers to stacks is decided so to minimize the total number of reshuffles required to retrieve all the containers in the yard, according to their exit times. Such a number is the optimal value of the *Block Relocation Problem* (BRP) associated to the yard. We will discuss such a problem in Sect. 2. Then, Function CF3 returns the optimal routes for the cranes that have to move containers among different stacks and for the trucks that enter (and then leave) the yard in order to retrieve a given set of containers.

In the rest of the paper, in Sect. 2 we will consider the Block Relocation Problem, whose solution leads the strategies defined to implement both CF1 and CF2, as described in Sects. 3.1 and 3.2 respectively. Finally, in Sect. 3.3, we will discuss the algorithms implemented for CF3.

2 The Block Relocation Problem

Denote by $\mathscr{S}$ a yard defined by a set $N = \{1, \ldots, n\}$ of containers allocated in w stacks of height (in terms of available slots/tiers) h. For each container i, it is given an *exit time* t_i that represents the time when i has to be retrieved from the yard. Without loss of generality, here and in the following, we will assume that the containers of N are ordered in such a way that $i < j$, implies $t_i \leq t_j$, for each $i, j \in \{1, \ldots, n\}$.

The Block (or Container) Relocation Problem (BRP) consists in deciding where to reallocate every container that is moved by a reshuffle operation, in order to minimize the total number of reshuffles needed to retrieve all the containers according to the retrieval order (1, …, n). Observe that such minimum is lower bounded by the number of *blocking* containers of $\mathscr{S}$, i.e., those located in any slot above a container with higher retrieval priority. Figure 3 gives an example of the BRP with a system defined by $w = 3$ stacks of height $h = 3$ and $n = 6$ containers. Starting from the initial yard, where the blocking containers are 5 and 6, the sequence of movements of an optimal solution is reported. At each step, the next container to be moved with a reshuffle or a retrieval operation is highlighted in gray. The minimum number of reshuffles required is three: container 6 is reshuffled in order to retrieve container 1; container 5 is reshuffled to retrieve container 2; we then reshuffle container 6 to retrieve container 4. Since BRP generalizes the Bounded Coloring Problem (also known as Mutual Exclusion Scheduling) on permutation graphs, it is known to be

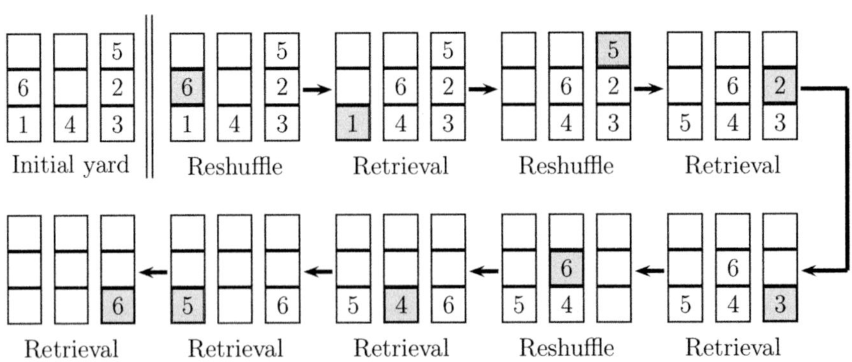

Fig. 3 A representation of an optimum solution for the Block Relocation Problem

NP-hard for any fixed $h \geq 6$. For some complexity results on BRP, Bounded Coloring and related problems, see [3, 6, 8, 11].

All exact approaches for the problem proposed in the literature for the BRP problem (see [10, 16, 19, 20]) cannot solve instances with more than 70 containers. Since this is far away from the realistic scenarios, where usually many hundreds (if not thousands) of containers are located in hundreds of stacks, a huge amount of heuristic procedures have also been proposed. Some effective procedures can be found in [5, 8, 12, 17, 20]. In particular, here (see also Sect. 3.1) we considered the Group Assignment Heuristic (GAH) introduced in [18], as it usually provides good solutions in a very reasonable amount of time. The algorithm is described in the following.

2.1 *The GAH Algorithm*

The GAH algorithm consists in an iterative procedure: at each iteration i, container i is retrieved from its stack, say $\bar{s}$, and the corresponding set B_i of n_i blocking containers that have to be reshuffled are assigned to other stacks of the yard according to a two phase criterion. We denote the containers of B_i as $c_1, \ldots, c_{n_i}$, according to a decreasing order of their retrieval times. Therefore, $c_k > c_{k+1}$ for any $k = 1, \ldots, n_i - 1$. Along the algorithm, for each stack s of the yard, we keep updated the value σ_s as the minimum retrieval order among all the containers located in s (σ_s is set to $n + 1$ if s is empty).

In the first phase of the assignment procedure, the blocking containers of B_i are considered one after the other from c_1 to c_{n_i}.

We define Λ_{c_k} as the set of stacks s such that: (i) s contains a free slot; (ii) $\sigma_s > c_k$; (iii) s does not contain any $c_{k'} > c_k$ such that $c_{k'}$ is located below c_k in the origin stack $\bar{s}$. If $\Lambda_{c_k} \neq \emptyset$, assign c_k to the stack $s_{c_k} \in \Lambda_{c_k}$ that minimizes σ_s. Otherwise, c_k is assigned in the second phase of the algorithm.

In the second phase, we iteratively consider, in the order $c_{n_i}, \ldots, c_1$ (and therefore by increasing retrieval time), all the containers of B_i that have not been assigned in the first phase. For any c_k in this set and for any stack $s \neq \bar{s}$, we define $\Omega_s^{c_k}$ as the set of containers $c_{k'} > c_k$ of B_i such that: (i) $c_{k'}$ has already been assigned to s; (ii) $c_{k'}$ is located below c_k in $\bar{s}$. Now, let $p_s^{c_k}$ be defined as follows:

$$p_s^{c_k} = \begin{cases} \sigma_s - c_k, & \text{if } \Omega_s^{c_k} = \emptyset \\ \min\{\sigma_s - c_k, c_k - c_{k'}\} & \text{if } \Omega_s^{c_k} = \{c_{k'}\} \\ 1 - n, & \text{if } |\Omega_s^{c_k}| > 1 \end{cases}$$

Then, c_k is assigned to the stack s with the minimum positive $p_s^{c_k}$, if any; otherwise, to the one with the maximum negative $p_s^{c_k}$. Ties are broken arbitrarily.

At the end of the two phases, all the containers of B_i are moved to the assigned stacks and the algorithm proceeds with considering the next container $i + 1$ to be retrieved.

3 Implementing the CONTRAST Functions

In the following we describe how we implemented the CONTRAST functions CF1, CF2, and CF3 introduced in Sect. 1.

3.1 The Function CF1: Reallocating the Reshuffled Containers

As mentioned before, function CF1 is called every time a container has to be retrieved from its current position in order to assign the containers that have to be reshuffled to the other stacks of the yard. In the old practice of the container terminal of Pomezia, this assignment was decided *on line* by the crane operators that, in most of the cases, just moved the containers to be reshuffled to the closest available stack, as they did not even know the exit times of the containers in the yard. We refer to this strategy as *CST*. Here, we solved such assignment problem by making use the algorithmic criterion introduced by [18] for the GAH heuristic, that we reimplemented in our code (strategy *GAH*).

In order to evaluate the performances of the two different procedures, we performed some computational experiments. The data set for these experiments has been produced by the research group of University of Genoa. In particular, they: (i) analyzed the historical dataset of the container terminal of Pomezia (referred to the years 2007–2012) and (ii) provided:

- the description of a container yard defined by a set of 308 stacks of height 3 or 4;
- 365 typical scenarios (one for each day of the year). Each scenario consists of about 600 (on average) containers, differently distributed in the stacks of the yard.

For each of these scenarios, we calculated the number of reshuffle operations needed to retrieve (according to their exit times) all the containers of the yard, following strategy *CST* (*GAH* resp.) for reallocating the reshuffled containers.

In Fig. 4 we resume the results of these computational experiments, where the line above (in black) is associated with the *CST* reallocation strategy, while the one below (in grey) indicates the values obtained by the *GAH* procedure. In the horizontal axis we represent the instances used in the experiments.

It is easy to see that (as expected) the *GAH* strategy outperforms the *CST* one. Indeed, the total number (over all the 365 instances) of reshuffle operations needed in the first case is 67.078 while, with the CST strategy, 118.589 reshuffles are needed. This means that the proposed solution saves, on average, about 56% of the reshuffled operations needed by the previous procedure.

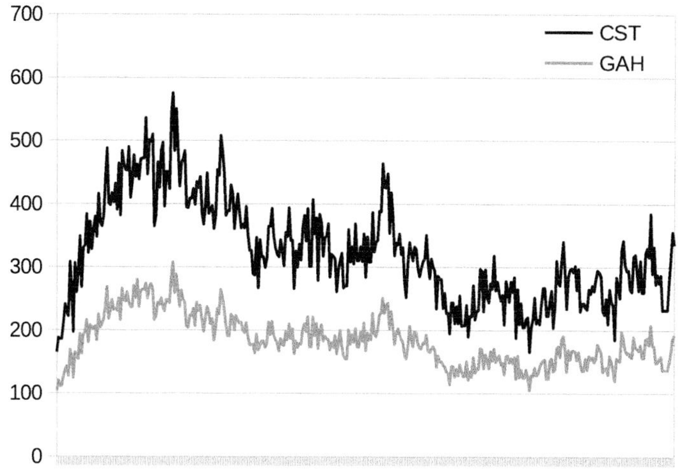

Fig. 4 Number of reshuffle operations needed to retrieve all the containers of the considered yard, according to the two strategies, *CST* and *GAH*, used to reallocate the reshuffled containers

3.2 *The Function CF2: Loading the Incoming Containers in the Yard*

When a train arrives at the terminal, the containers on the train must be removed and stored in the stacks of the yard. As observed in Sect. 2, any blocking container of a yard will be reshuffled at some point to retrieve some container with an higher exit priority. Therefore, the incoming containers are assigned to the stacks of the yard so to minimize the number of those becoming blocking. Recall that such containers can be withdrawn from the train in any order, since they are not piled in stacks. Therefore we can model the assignment problem as follows.

Let B the set of the incoming containers and let S be a yard. Moreover, denote by P the stacks of the yard with empty slots. For any $c \in B$, we are given a type t_c and we denote by P_C the stacks of P that can accommodate containers of type t_c. Moreover, for any $u \in P$ let q_u be the number of available slots of stack u (available capacity). For each container $c \in B$ and each stack $u \in P_c$, we introduce a binary variable $x_{c,u}$ that takes value 1 if c is assigned to u and 0 otherwise. The cost w_{cu} of this assignment is 0 if stack u does not contain containers with retrieval priority higher than the one of c, while $w_{uv} = 1$ otherwise (i.e. if c is blocking once allocated in u). We want to compute an assignment of containers to stacks so that: (a) all containers of B are allocated in some available stack, (b) the number of containers assigned to each stack u does not exceed q_u, (c) the total cost of the allocation is minimized. This problem is a special case of the Generalized Assignment Problem (GAP) and it can be formulated as an integer program:

$$\min \sum_{c \in B} \sum_{u \in P_c} w_{cu} x_{cu} \tag{1}$$

$$\sum_{u \in P_c} x_{cu} = 1 \qquad c \in B \tag{2}$$

$$\sum_{c|u \in P_c} x_{cu} \leq q_u \qquad u \in P \tag{3}$$

$$x_{cu} \in \{0, 1\} \qquad c \in B, u \in P_c \tag{4}$$

The constraints guarantee, respectively, that each container is assigned to one and only one stack and that the total number of containers allocated to each stack does not exceed its available capacity. In its general version, GAP has weights and profits (depending on items and bins) and it is known to be an NP-hard problem. On the contrary, in our special case, all the items weights are identical and therefore the problem can be solved as an s-t min-cost flow problem (see [1]) on a suitable graph $G = (N, A)$. In particular, the set of nodes N is defined by $B \cup P \cup \{s, t\}$. Then, for each $c \in B$ and $u \in P_c$, there exists an arc $(c, u) \in A$ with cost w_{cu} and capacity 1. Moreover, for any $c \in B$, we have an arc (s, c) with cost 0 and capacity 1 and, for any $u \in P$, an arc (u, t) with cost 0 and capacity q_u. Each s-t flow of value $|B|$ on such graph corresponds to a feasible assignment of the blocks of B to the stacks of P. In function CF3, we did use a standard C++ implementation for the s-t min-cost flow algorithm provided by the library Lemon [9]. For a dedicated approach to solve this special GAP without using a min-cost flow, see [4].

As for function CF1, we also made some computational experiments to evaluate the performances of the procedure proposed for CF2. Also here, we used the dataset of 365 instances introduced in Sect. 3.1. In particular, for each of these instances, we randomly generated 20 containers that are supposed to arrive by a train and need to be allocated in the available stacks of the yard. Then we took in to account two methods to decide such allocation: (a) for the same reasons explained in the previous section, we considered the case where each incoming container is assigned to the available stack of the yard that is the closest to the entry platform (strategy *CST*); (b) according to strategy *minBLOCK*, the assignment problem of the incoming containers is solved by the CF2 function described above. Once the incoming containers are allocated in the stacks of the yard, we apply the GAH procedure in order to count the reshuffle operations needed to empty the terminal.

The results of these computational experiments are resumed in Table 1.

Such results give evidence of the relevance of a good strategy to allocate the containers that enters the terminal. In particular, over all the 365 considered instances, one needs 67,227 reshuffle operations to empty the yard after *minBLOCK* strategy is applied, while such a number increases to 71,820 if one uses *CST*. Therefore, the first procedure allows to save (on average) 0.63 reshuffle operations for each incoming container.

Table 1 Here we present the computational results obtained by emptying the yard, once 20 incoming containers are allocated in the stacks of the terminal according to the two strategies CST and minBLOCK, respectively. The Table reports the total and average number of reshuffle operations for all the 365 instances taken under consideration

CST		minBLOCK	
Total	Average	Total	Average
71,820	196.77	67,227	184.18

3.3 *Moving the Containers Within the Yard*

When a container must be moved by a crane between two given stacks or enter/exit platforms of the yard, the route of the crane should correspond to the shortest path between these two endpoints. Indeed, we calculate off-line the shortest path for each couple of stacks and platforms of the yard, by a standard implementation of the Dijkstra algorithm [15].

More interesting is the case when a truck enters the terminal in order to collect a set C of containers that have to be retrieved. Let S be the set of stacks in which such containers are allocated. Since the retrieval order is not given in advance, we assign the truck a route that corresponds to a minimum distance Hamiltonian cycle among the stacks of S. This cycle is calculated by a standard implementation of the well known dynamic programming algorithm for the Traveling Salesman Problem (see [13]).

Acknowledgements We would like to thank Paolo Dell'Olmo and Umberto Ferraro Petrillo (University "La Sapienza" of Rome) and Anna Sciomachen (University of Genoa) for their fundamental contribution to the CONTRAST project.

This paper has received funding from the European Unions Horizon 2020 research and innovation programme under the Marie Skodowska-Curie grant agreement No 764759.

References

1. Ahuja, R.K., Magnanti, T.L., Orlin, J.B.: Network Flows: Theory, Algorithms, and Applications. Prentice Hall Inc., Englewood Cliffs (1993)
2. Ambrosino, D., Sciomachen, A.: How to reduce the impact of container flows generated by a maritime terminal on urban transport. In: Brebbia, C.A. (ed.) Sustainability Today, pp. 79–88 (2012)
3. Bacci, T., Mattia, S., Ventura, P.: Some complexity results for the minimum blocking items problem. In: Optimization and Decision Science (ODS 2017): Methodologies and Applications. Springer Proceedings in Mathematics and Statistics, vol. 217, pp. 475–483 (2017)
4. Bacci, T., Mattia, S., Ventura, P.: A new lower bound for the block relocation problem. In: International Conference on Computation and Logistics – ICCL 2018. Lecture Notes in Computer Science, vol. 11184, pp. 168–174 (2018)
5. Bacci, T., Mattia, S., Ventura, P.: The bounded beam search algorithm for the block relocation problem. Comput. Oper. Res. **103**, 252–264 (2019)

6. Bonomo, F., Mattia, S., Oriolo, G.: Bounded coloring of co-comparability graphs and the pickup and delivery tour combination problem. Theor. Comput. Sci. **412**(45), 6261–6268 (2011)
7. Carlo, H., Vis, I., Roodbergen, K.: Storage yard operations in container terminals: literature overview, trends, and research directions. Eur. J. Oper. Res. **235**(2), 412–430 (2014)
8. Caserta, M., Schwarze, S., Voß, S.: A mathematical formulation and complexity considerations for the blocks relocation problem. Eur. J. Oper. Res. **219**, 96–104 (2012)
9. COIN-OR: The lemon library. https://lemon.cs.elte.hu/trac/lemon (2013)
10. Galle, V., Barnhart, C., Jaillet, P.: A new binary formulation of the restricted container relocation problem based on a binary encoding of configurations. Eur. J. Oper. Res. **267**, 467–477 (2018)
11. Jansen, K.: The mutual exclusion scheduling problem for permutation and comparability graphs. Inf. Comput. **180**, 71–81 (2003)
12. Jovanovic, R., Voß, S.: A chain heuristic for the blocks relocation problem. Comput. Ind. Eng. **75**, 79–86 (2014)
13. Lawler, E.L., Lenstra, J.K., Kan, A.H.G.R., Shmoys, D.B. (eds.): The Traveling Salesman Problem: A Guided Tour of Combinatorial Optimization. Wiley-Interscience, New York (1985)
14. Lehnfeld, J., Knust, S.: Loading, unloading and premarshalling of stacks in storage areas: survey and classification. Eur. J. Oper. Res. **239**, 297–312 (2014)
15. Schrijver, A.: Combinatorial Optimization: Polyhedra and Efficiency. Springer, Berlin (2003)
16. Tanaka, S., Takii, K.: A faster branch-and-bound algorithm for the block relocation problem. IEEE Trans. Autom. Sci. Eng. **13**(1), 181–190 (2016)
17. Unluyurt, T., Aydin, C.: Improved rehandling strategies for the container retrieval process. J. Adv. Transp. **46**(4), 378–393 (2012)
18. Wu, K.C., Ting, C.J.: Heuristic approaches for minimizing reshuffle operations at container yard. In: Asia Pacific Industrial Engineering & Management Systems Conference (2012)
19. Zehendner, E., Caserta, M., Feillet, D., Schwarze, S., Voß, S.: An improved mathematical formulation for the blocks relocation problem. Eur. J. Oper. Res. **245**, 415–422 (2015)
20. Zhu, W., Qin, H., Lim, A., Zhang, H.: Iterative deepening A* algorithms for the container relocation problem. IEEE Trans. Autom. Sci. Eng. **9**, 710–722 (2012)

Product Design

Design Optimization of Synchronous Reluctance Motor for Low Torque Ripple

Andrea Credo, Andrea Cristofari, Stefano Lucidi, Francesco Rinaldi, Francesco Romito, Marco Santececca and Marco Villani

Abstract The aim of this paper is to optimize the design of multiple flux barriers *Synchronous Reluctance Motor* in order to smooth the torque profile without rotor skewing. A new strategy is proposed by modelling the particular optimal design problem as mixed integer constrained minimization of a suitable objective function. The procedure has allowed to optimize the rotor shape for minimum torque ripple starting from an existing stator core.

Keywords Synchronous reluctance · Torque ripple optimization · Finite element analysis · Derivative-free algorithms · Nonlinear mixed-integer optimization methods

---

A. Credo · M. Santececca · M. Villani
Dipartimento di Ingegneria Industriale e dell'Informazione e di Economia, Università degli studi dell'Aquila, Via Giovanni Gronchi, 18, 67100 L'Aquila, Italy
e-mail: andreacredo.93@gmail.com

M. Santececca
e-mail: msante@hotmail.it

M. Villani
e-mail: marco.villani@ing.univaq.it

A. Cristofari · F. Rinaldi
Dipartimento di Matematica, Università degli studi di Padova, Via Trieste, 63, 35121 Padua, Italy
e-mail: andrea.cristofari@unipd.it

F. Rinaldi
e-mail: rinaldi@math.unipd.it

S. Lucidi (✉) · F. Romito
Dipartimento di Ingegneria Informatica, Automatica e Gestionale, Sapienza Università di Roma, Via Ariosto, 25, 00185 Rome, Italy
e-mail: lucidi@diag.uniroma1.it

F. Romito
e-mail: romito@diag.uniroma1.it

M. Dell'Amico et al. (eds.), *A View of Operations Research Applications in Italy, 2018*, AIRO Springer Series 2, https://doi.org/10.1007/978-3-030-25842-9_5

1 Introduction

Synchronous Reluctance Motors (SRM) could be considered as alternative to its counterparts, namely permanent magnet and induction motors and have advantages such as low cost and high efficiency. The transversally laminated rotor is preferable and exhibits a reduced inertia due to air-flux barriers (Fig. 1).

The torque generated by SRM depends on the magnetic saliency of the rotor: no windings, brush or permanent magnets (PM) are used which makes the SRM a simple and robust electric machine.

Compared with permanent magnet machines, the SRM can operate at high speed because of the easier field weakening capability and the rugged rotor. If compared with induction motors it has theoretically no rotor losses and a comparable torque density depending on the design of the machine; moreover, the SRM has lower temperature rise in the rotor and shaft and consequently good efficiency and higher reliability.

On the other hand, synchronous reluctance machines usually present some additional issues like lower power factor but above all higher torque ripple that is intolerable in *most applications* [6, 12, 17]: it is due to the interaction between the spatial harmonics of electrical loading and the *rotor anisotropy,* and gives rise to noise and vibrations.

The easiest and commonly adopted solution for the torque ripple reduction is the skewing of the rotor stack (Fig. 2). It could appear as a simple and cheap measure to actuate but gives rise to some drawbacks. Particularly, a slight reduction of the output torque and then a difficulty in the insertion of permanent magnets inside the rotor core for the low cost PM-assisted SRM [1]. In this case, the continuous rotor skewing cannot be adopted and the alternative solution is a multi-stack rotor core where the rotor is split into two or more parts, each of them is skewed with respect to the others: this solution could be too expensive and not convenient to adopt from the industrialization point of view.

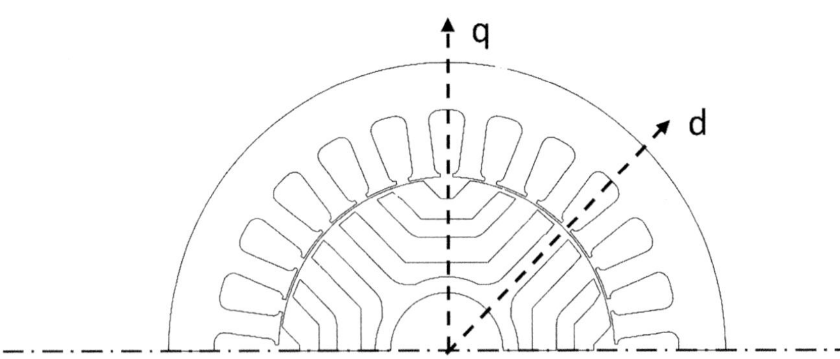

Fig. 1 Cross section of SRM with transversally laminated rotor (half machine)

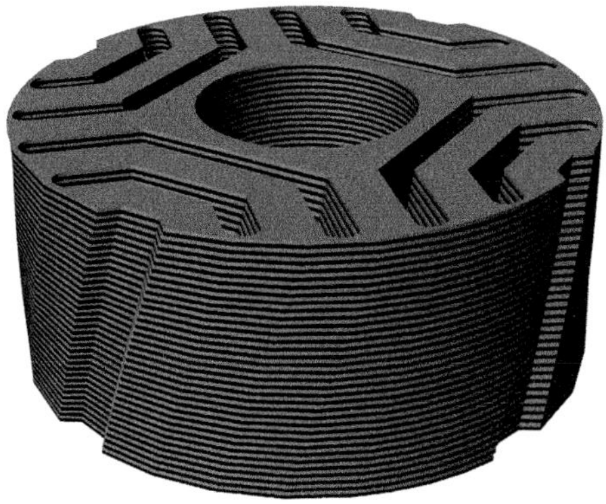

Fig. 2 3D view of skewed rotor

The design of the SRM requires a fine analysis and an accurate prediction of machine parameters and performances [2]. The reduction of torque ripple represents one of the most relevant constraints during the design step and the aim is to find a solution that allows to satisfy the requirements and simplify the motor assembling avoiding the skewing. This can be achieved by an appropriate combination of design parameters and a right choice of the number of flux barriers with respect to the stator slots.

This paper presents a novel strategy to reduce the torque ripple of the synchronous reluctance motor by an *optimization algorithm recently proposed in* [7]. The focus is to smooth the torque profile by varying the dimensions of the rotor shape only without any change in the stator core and winding and above all without rotor skewing. For this, a commercial three-phase induction motor has been chosen substituting the squirrel cage rotor with a multiple flux-barriers one. The new rotor has been designed and then refined by the optimization algorithm in order to reduce the torque ripple.

2 The Preliminary Design of the SRM

In this step, the preliminary design of the SRM has been carried out starting from a commercial three-phase induction motor (3 kW, 4 pole, 400 V, 50 Hz, TEFC, IEC size 100): Table 1 presents the motor requirements.

The same stator core and housing has been adopted, substituting the squirrel cage rotor with a flux-barriers one: this approach allows to minimize the investment costs in new tooling using existent components.

The design of SRM is mainly focused on the rotor creation and, more specifically, the type of flux barrier topology. For this, a sizing procedure has been used for the design of the rotor shape: it allows to determine the rotor dimensions (Fig. 3) by adopting simplified relationships between geometrical and physical motor data in order to meet the specifications. These data allow to calculate the motor performance i.e. the axis fluxes, the magnetic potential differences, the flux densities, the axis currents, the power factor and the torque. A detailed description of the sizing procedure is given in [13, 14].

Table 1 Motor requirements

Rated power	kW	3.0
Rated speed	rpm	1500
Rated torque	Nm	19
Number of poles		4
Number of slots		36
Outer stator diameter	mm	150
Inner stator diameter	mm	90
Stack length	mm	150
Airgap length	mm	0.50
Electrical steel		M470-50A
Cooling		Air cooled

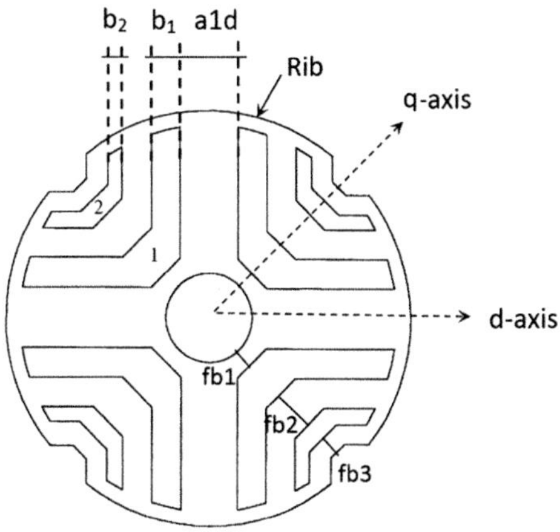

Fig. 3 Rotor dimensions

The sizing of rotor is not easy to carry out due to presence of the flux barriers whose number and dimensions affect heavily the *saliency ratio*, that is the ratio between d and q-axis inductances. For this reason, the authors have introduced several relationships among the rotor dimensions (Fig. 3) and the rotor diameter (D_r), by means of suitable coefficients (k_i).

These relations and the coefficients have been refined and tested analyzing several SRMs, of different size, by the Finite Element analysis. Therefore, these relations represent a guideline for the designer and *should orient her/him towards* the search of a good solution. For a rotor with two flux barriers they are (Fig. 3):

$$k_1 D_r \le fb1 \le k_2 D_r,$$
$$k_3 D_r \le a1d \le k_4 D_r,$$
$$k_5 D_r \le b_1 \le k_6 D_r,$$
$$k_7 D_r \le b_2 \le k_8 D_r,$$

$$fb_2 \geq k_9 fb1 \quad \text{and} \quad fb_3 \geq k_{10} D_r. \tag{1}$$

This approach will be very useful for the next optimization step. In fact, the coefficients k_i are imposed in the Input data and these values define a range within each variable should vary according to the imposed values on D_r. The sizing process starts considering, for each variable, the average value and this choice allows to obtain a preliminary rotor design that certainly does not represent the final design. *Design refinement is obtained* by the optimization algorithm (Sect. 4).

The number of flux barriers (N_{barr}) has been calculated according to the number of stator slots (N_{slot}) and the following relationship:

$$\left(N_{\text{barr}}\right)_{\text{TOT}} = N_{\text{pole}}\ K_s, \tag{2}$$

where N_{pole} is the number of poles and

$$K_s \neq \frac{N_{\text{slot}}}{N_{\text{pole}}}. \tag{3}$$

It is important to underline that increasing the number of fluxbarriers, the torque tends to increase whereas the torque ripple decreases.

Since the chosen stator core presents 36 slots, 24 barriers have been chosen according to (2), (3) and particularly 6 barriers per pole: this choice has also been conditioned by the constraint on the outer rotor diameter.

The cross section of the preliminary design is shown in Fig. 4. This configuration is a quite conventional reluctance geometry, characterized by three flux barriers that share the same width, tilt angle and distance between them. Also the barrier fillet radius at the external rib is the same. An external channel located along the q-axis has been considered, in order to simplify the welding procedure of the rotor stack.

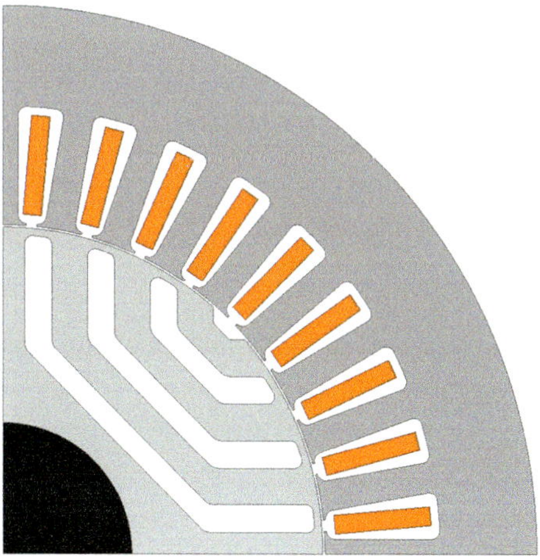

Fig. 4 Cross-section of the preliminary design (1 pole)

About the thickness of ribs (Fig. 3) they should be chosen according to the improvement of saliency ratio and the mechanical strength. From a preliminary analysis a minimum value of 1 mm has been imposed: this value is consistent with the maximum speed and mechanical stress.

The preliminary design has been analyzed by the Finite Element (FE) software, since the SRM presents notable non-linear characteristics due to the effects of the saturation and cross-coupling phenomena. An accurate bi-dimensional FE model has been developed introducing a parametric model in order to modify any geometric dimensions of the rotor shape.

A transient analysis has been carried out and one pole has been simulated thanks to the motor symmetry.

The behavior of the electromagnetic torque, as a function of the rotor position, has been calculated by a series of time-based simulations: then its average value along the entire electrical period has been evaluated. More in the detail, it can be proved that for the considered combination $N_{\text{slot}}/N_{\text{barr}}$, the electrical period of the torque is equal to 60 *electrical degrees (30 mechanical degrees).* In turn, each transient analysis consists of 80 computational steps, which means that the electromagnetic torque is represented with an accuracy of 0.75 *electrical degrees.*

We performed the time stepping FE simulations (at 1500 rpm) by imposing the axis currents in order to maximize the torque-current ratio: moreover, a conventional temperature of 75 °C has been chosen for the stator winding. Table 2 presents the main data and performance of the preliminary design. The torque ripple, defined *as the ratio* between the difference of the maximum and minimum values of torque and the average one is about 20% for the no-skewed rotor. Moreover, a further analysis has been carried out with a skewed rotor in order to verify the effect on the torque behavior: a continuous rotor skewing has been simulated with the skewing angle equal to the stator slot angle. Figure 5 presents the preliminary torque profile and new one: in the latter case, the average torque is 18.3 Nm and the ripple 8.1% with a reduction of *about 60% with respect* to the no-skewed rotor.

3 Design Optimization of the SRM

The demand of high-performance motors requires the definition of more efficient designs. This aim can be achieved by interfacing the Finite Element software with automatic optimization procedures [7, 8, 15].

In particular the FE analysis can be used to evaluate the motor performance, namely to compute the objective function value and some of the constraints. The optimization procedure can use the information obtained by the FE program to iteratively update the set of motor parameters and try to identify an optimal motor by making a trade-off between the different parameters of the machine.

Table 2 Motor requirements

Outer stator diameter	mm	150
Inner stator diameter	mm	90
Stack length	mm	150
N.turns per phase		180
Wire size	mm^2	1.135
N.barriers per pole		6
Width of the rotor tooth along d-axis ($a1d$)	mm	6
Thickness of rotor ribs	mm	1
Width of barrier 1 (b_1)	mm	3
Width of barrier 2 (b_2)	mm	3
Width of barrier 3 (b_3)	mm	3
Performance		
Phase voltage	Vrms	220
Phase currents	Arms	6.5
Average torque	Nm	19.3
Base speed	rpm	1500
Phase resistance @ 75 °C	ω	1.78
Joule losses	W	226
Iron losses	W	81
Efficiency	%	90.1
Torque ripple (no-skewed)	%	20.1

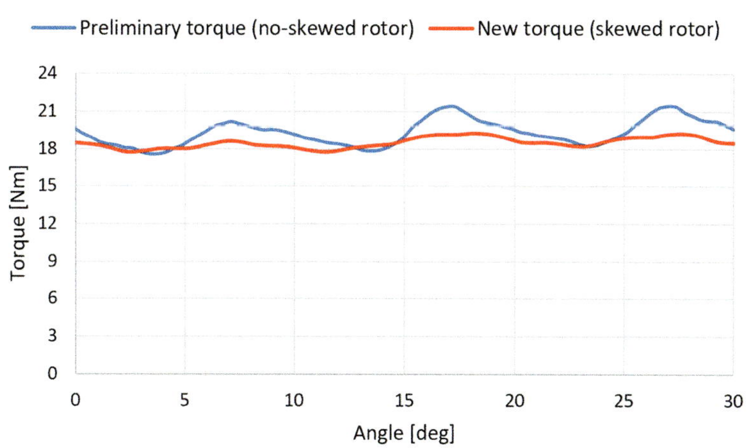

Fig. 5 Torque behavior: preliminary design (no-skewed rotor) and new design (skewed rotor)

The set of parameters x used in the optimization procedure of the SRM concern the rotor shape only and are listed in Table 3. *The tilt angles (alfa and beta) define the terminal shape of the flux barriers: if these angles are zero, the barriers assume the shapes shown in Fig.* 3. Figure 6 shows in details the 17 design variables (dimensions of flux barriers, tilt angles, fillet radii): they have been varied according to practical limits existing in cutting the rotor shape.

Table 3 Minimum and maximum ranges of design variables

Variables		Min	Max	Type	Step
x_1. Width of rotor tooth along d-axis (mm)	$a1d$	4	8	$\mathbb{R}$	–
x_2. Width of barrier 1	b_1	2	4.5	$\mathbb{R}$	–
x_3. Width of barrier 2	b_2	2	4.5	$\mathbb{R}$	–
x_4. Width of barrier 3	b_3	2	4.5	$\mathbb{R}$	–
x_5. Distance between shaft and barrier 1	$fb1$	2	3	$\mathbb{R}$	–
x_6. Distance between barriers 1–2	$fb2$	3	5	$\mathbb{R}$	–
x_7. Distance between barriers 2–3	$fb3$	3	5	$\mathbb{R}$	–
x_8. Distance between barriers 3-ext.channel	$fb4$	3	5	$\mathbb{R}$	–
x_9–x_{10}. Tilt angles barrier 1	α_1, β_1	0°	5°	$\mathbb{R}$	–
x_{11}–x_{12}. Tilt angles barrier 2	α_2, β_2	0°	5°	$\mathbb{R}$	–
x_{13}–x_{14}. Tilt angles barrier 3	α_3, β_3	0°	5°	$\mathbb{R}$	–
x_{15}. *Number of conductors in slot*		*20*	*40*	$\mathbb{Z}$	*1*
x_{16}–x_{17}. Fillet radii (mm)	ri, re	0.5	3	$\mathbb{R}$	–

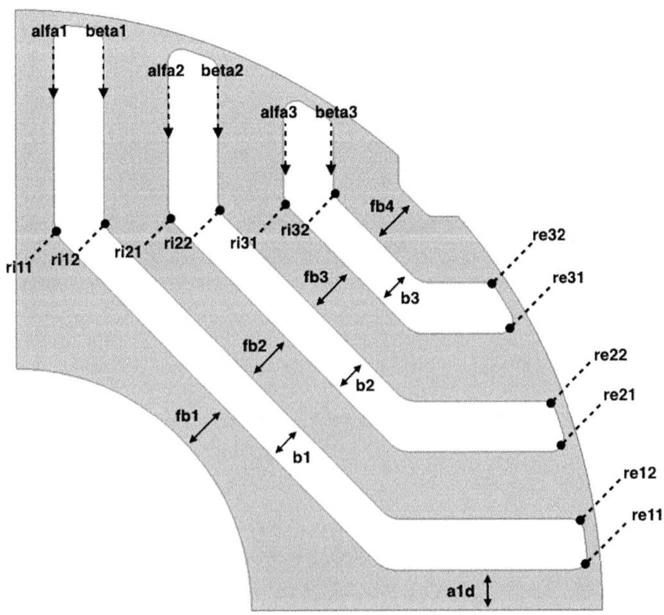

Fig. 6 Rotor independent variables

Table 4 Constraints

Constraints		Limits
c1. Fill factor		*≤0.48*
c2. Max flux density in the stator tooth	T	≤1.60
c3. Max flux density in the stator yoke	T	≤1.45
c4. Max flux density in the rotor tooth	T	≤1.50
c5. Joule losses at 90 °C	*W*	*≤240*
c6. Torque ripple	%	≤10.0
c7. Average torque @ 1500 rpm	Nm	≥19.0

Moreover, several constraints have been introduced to guarantee the reliability and feasibility of the final design (Table 4). They are: the average torque, the maximum flux densities in the stator and rotor core and the percentage of torque ripple.

The aim of the optimization was to refine the rotor shape, without skewing, in order to smooth the torque profile in the electrical period. This approach has required many FE calls for each optimization step and high computation time, but represents a powerful tool for an accurate design of the machine.

4 The Optimization Algorithm

The above described design problem can be modeled by an optimization problem, where the functions describing the motor behavior with respect to its parameters are unknown and they can be only computed by complex simulation programs. So, the derivatives of these functions are unknown as well, and then, a derivative-free algorithm must be employed.

A further difficulty arising in our context is the high computational burden needed to evaluate the motor behavior in the entire electrical period.

Moreover, the aim of the minimization procedure is twofold: on the one hand, the torque profile must be smoothed, and, on the other hand, its average value must be maximized. So, a multiobjective problem could be in principle introduced.

Several multiobjective derivative-free methods have been proposed in the literature (see e.g. [3, 10] and references therein for further details). Some of those methods take into account preferences of the decision maker (i.e. the person that can give preference information concerning the solutions) relating the objectives.

As we said, in the considered context there is usually the need for a trade-off between a high expected value of the torque profile and a small torque profile variance.

For any motor parameter vector x, we indicate by $a(x)$ and $\sigma(x)$ the expected value and the standard deviation of the torque profile in the electrical period, respectively. One way we have to properly take standard deviation into account is to minimize a proper combination of the two objectives, that is,

$$f(x) = -a(x) + \gamma\sigma(x),$$

thus obtaining a "risk-sensitive" objective function.

The parameter $\gamma \geq 0$ is called *risk-aversion parameter*, since it sets the relative values of torque profile standard deviation and expected value (for $\gamma > 0$, we are willing to trade off a decrease in the expected value of the torque profile for a sufficiently large decrease in the torque profile standard deviation). In our case, we set the risk-aversion parameter $\gamma = 10^4\sqrt{N}$, where N is the number of measures in the electrical period (i.e., $N = 61$). In particular, we deal with the following mixed integer non-linear minimization problem:

$$\begin{aligned} & \min\ f(x) \\ & g(x) \leq 0 \\ & l \leq x \leq u \\ & x_i \in \mathbb{Z}, \quad i \in I_z \\ & x_i \in \mathbb{R}, \quad i \in I_c = \{1, \dots, n\} \setminus I_z, \end{aligned} \tag{4}$$

where $f : \mathbb{R}^n \rightarrow \mathbb{R}$, $g : \mathbb{R}^n \rightarrow \mathbb{R}^m$, $x \in \mathbb{R}^n$, $l \in \mathbb{R}^n$, $u \in \mathbb{R}^n$, I_z is the index set of the integer variables and I_c is the index set of the continuous variables.

In problem (4), x is the motor parameter vector and the constraints are those described in the previous section.

Recalling the considerations made above on the computational burden needed to evaluate the objective function, *the employment of a global optimization algorithm* (e.g. [11, 16]) *can be impractical, due to the scarce efficiency of such algorithms when only a limited number of function evaluations is available.* So, we employ a local derivative-free algorithm to solve problem (4). In particular, we use *Algorithm DFL*, recently proposed in [9], which is based on a sequential penalty approach. Namely, the non-linear constrained problem (4) is approximated by a sequence of box-constrained problems of the form:

$$\begin{aligned} & \min\ P(x; \varepsilon_k) = f(x) + \frac{1}{\varepsilon_k} \sum_{i=1}^{m} \max\{0, g_i(x)\}^p \\ & l \leq x \leq u \\ & x_i \in \mathbb{Z}, \quad i \in I_z \\ & x_i \in \mathbb{R}, \quad i \in I_c, \end{aligned}$$

where $\varepsilon_k > 0$ and $p > 1$.

Algorithm DFL generates a sequence $\{\varepsilon_k\}$ such that, for any given ε_k, function $P(x; \varepsilon_k)$ is reduced (with respect to x) by means of suitable derivative-free line searches along all coordinate directions. Afterwards, if the constraint violation at the new point is not sufficiently decreased, then the penalty parameter ε_k is updated. We report in Fig. 7 a scheme of Algorithm DFL, highlighting the main steps of the

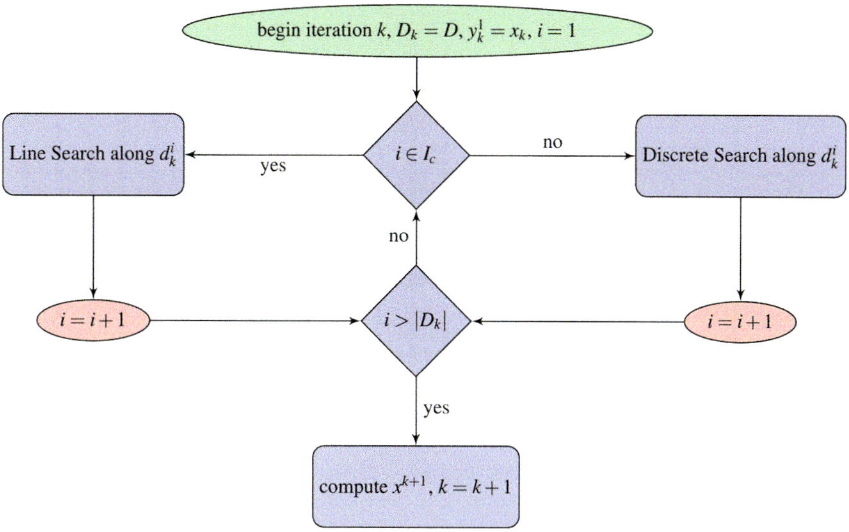

Fig. 7 Algorithm DFL

method. A more detailed description of the algorithm and the analysis of its theoretical properties can be found in [9]. Moreover, a Fortran 90 code of Algorithm DFL is freely available for download from the web page http://www.iasi.cnr.it/~liuzzi/DFL.

As we can see by taking a look at the scheme, the method is based on a suitable sampling strategy along a specific set of search directions $D = \{\pm e_i\}$, where e_i is the vector made of all zeros except for the ith entry that is equal to 1, for all $i = 1, \ldots, n$. So, by exploring a given direction, we move in a component-wise fashion.

Hence, we can use a specific search strategy (Continuous/Discrete) depending on the direction (component) we choose. When moving a continuous variable, the goal is guaranteeing both a sufficiently large movement along the search direction and a sufficient decrease in the objective function. When dealing with discrete variables, we further want to sample the objective function in such a way that integrality is always satisfied. At each iteration, after all directions have been explored, we check if an updating of the penalty parameter is needed and, finally, we update the iterate.

The sampling strategy is able to get, in the limit, sufficient knowledge of the problem functions (by using the Continuous and Discrete search) to recover both first order information for the continuous variables and some sort of local optimality for the discrete ones.

For what concerns the termination criterion, *we stopped the algorithm when the maximum value of the trial stepsize for the continuous variables is less than or equal to* 10^{-6}.

Now, we show the numerical results achieved by Algorithm DFL. The number of iterations and function evaluations required to solve the problem are equal to 1000

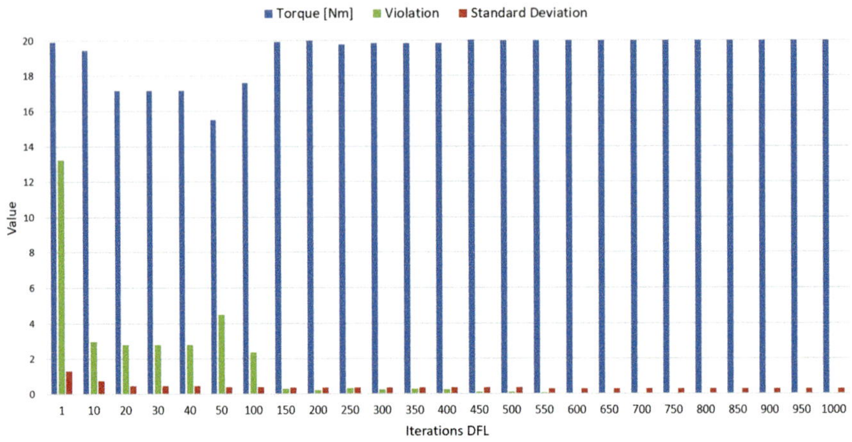

Fig. 8 Behavior of the average torque, standard deviation and constraint violation during the minimization by Algorithm DFL

Table 5 Comparison between DFL and CMA-ES

	Average torque	Standard deviation	Constraint violation
DFL	20	0.30	$4 \cdot 10^{-2}$
CMA-ES	5.33	0.13	14.74

and 1697, respectively. The final value of the torque is 20.00 with a standard deviation 0.30 and a constraint violation of $4 \cdot 10^{-2}$.

In Fig. 8, we focus on its computational behavior by reporting the value of the average torque, the standard deviation of the torque profile and the constraint violation (i.e., $\sum_{i=1}^{m} \max\{0, g_i(x^k)\}$) versus iteration k, respectively.

More in detail, the figures show that after 120 iterations (corresponding to 208 function evaluations) the algorithm is able to find an efficient solution, with an average torque equal to 19.89, a standard deviation of torque profile equal to 0.38 and a constraint violation equal to 0.32.

Finally, to evaluate the possible limits and drawbacks of using a local algorithm, we also tried a global algorithm. In particular, we considered an extension of CMA-ES Algorithm for mixed integer problems [4, 5], downloaded from http://cma.gforge.inria.fr. Default parameters were used and the algorithm was stopped after 10000 function evaluations. The inequality constraints $g_i(x) \leq 0$ were handled as in Algorithm DFL, that is, they were added to the objective function as a penalty term weighted by a parameter ε_k, updated as described above for Algorithm DFL.

In Table 5, we report the final results achieved by CMA-ES, compared with those obtained by DFL.

Table 5 seems to indicate that the global strategy implemented in CMA-ES needs a large number of function evaluations to locate a satisfying solution, especially for what concerns feasibility. *Indeed, CMA-ES quickly finds a solution with a small torque profile variance but a large constraint violation and then it spends the remaining budget of function evaluations without an improvement in feasibility.* Moreover, the results reported in the table show that, even with a limited budget of evaluations, the DFL method is able to get good points both in terms of objective function values and feasibility. The comparison with CMA-ES further indicates that the way DFL handles constraints and integer variables is efficient and, in this specific context, guarantees much better performances.

5 Results

The main data of the optimized design are shown in Table 6: it includes some of the key machine dimensions and performance at 1500 rpm. The same phase current of previous analysis has been imposed. *This design satisfies the constraints with sufficient accuracy.*

The cross-section of the new rotor is shown in Fig. 9. The comparison with the preliminary design (Fig. 4) points out a difference on the flux barriers dimensions and a slight weight reduction of rotor core (about 7%): this allows to reduce the inertia of the machine that could be very useful if high dynamic performance are required.

The torque profile (no-skewed rotor) is shown in Fig. 10 and the ripple is 6.67%: this behavior can be compared with the red curve shown in Fig. 5 (skewed rotor). Moreover, the axis inductances have been determined and in Fig. 11 are shown the profiles: from these curve the average saliency ratio has been evaluated and it is about 6.4.

Figure 12 presents the flux density distribution where the maximum values in the stator teeth and yoke are fully consistent with the imposed constraints (Table 6).

In Table 7 are compared the main performance of the proposed solutions. The torque ripple of the optimized design is lower than the preliminary design one with skewed rotor, and the average torque is slightly higher. Respect to the preliminary design without skewing, the new motor presents a significant ripple reduction of about 67%.

A further analysis has been carried out to determine the flux weakening capability of the optimized design. Figure 13 shows the torque versus speed of the machine.

All these results are quite satisfactory and confirm the goodness of the optimization procedure and the effectiveness of the proposed approach.

The aim was to smooth the torque profile by varying the dimensions of the rotor shape only without any change in the stator core and winding and above all without rotor skewing.

The results have demonstrated that this aim has been fully reached. The new design is a simple and cheap solution to industrialize and it is suitable for the insertion of proper amount of permanent magnets in the flux barriers for the low cost PM-assisted SRMs.

Table 6 Main data and performance of the optimized design

Outer stator diameter	mm	150
Inner stator diameter	mm	90
Stack length	mm	150
N. barriers per pole		6
Width of the rotor tooth along d-axis ($a1d$)	mm	5.78
Thickness of rotor ribs	mm	1
Width of barrier 1 ($b1$)	mm	3.06
Width of barrier 2 ($b2$)	mm	3.16
Width of barrier 3 ($b3$)	mm	4.46
Distance between shaft and barrier 1 ($fb1$)	mm	2.19
Distance between barriers 1–2 ($fb2$)	mm	4.16
Distance between barriers 2–3 ($fb3$)	mm	4.08
Performance		
Phase voltage	Vrms	220
Phase currents	Arms	6.5
Current angle	°	65
Average torque	Nm	20
Rated speed	rpm	1500
Phase resistance @ 75 °C	ω	1.78
Joule losses	W	226
Iron losses	W	78
Efficiency	%	90.0
Power factor		0.81
Torque ripple (no-skewed)	%	6.67

Fig. 9 Optimized design: cross section of rotor shape (1 pole)

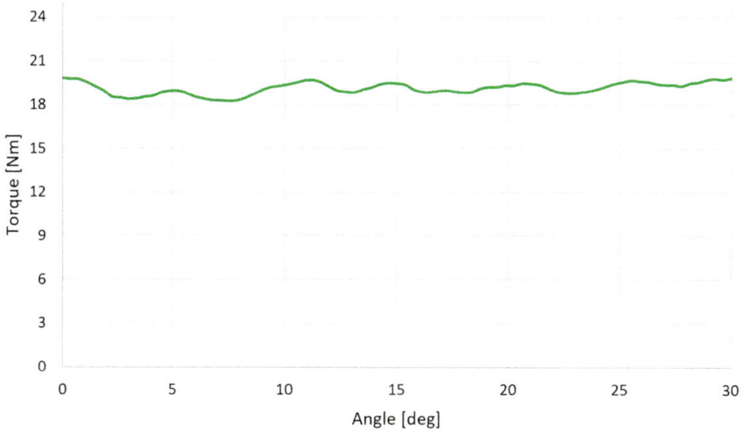

Fig. 10 Optimized design: torque behavior (no-skewed)

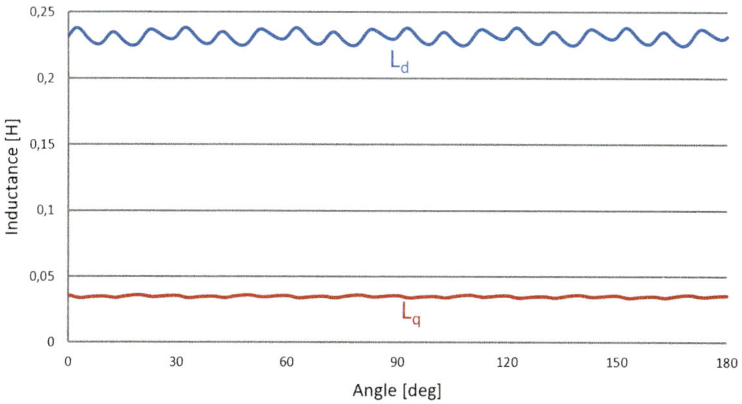

Fig. 11 Optimized design: d-q axis inductance values versus rotor position (6.5 Arms)

Fig. 12 Optimized design: flux density (6.5 Arms)

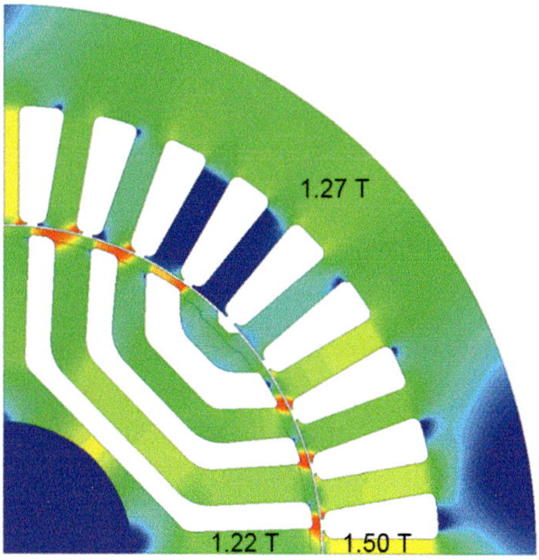

Table 7 Performance of the proposed solutions

		Preliminary		Optimized
		No-skewed	Skewed	Skewed
Phase current	Arms	6.5	6.5	6.5
Average torque	Nm	19.3	18.3	20
Torque ripple	%	20.1	8.1	6.67

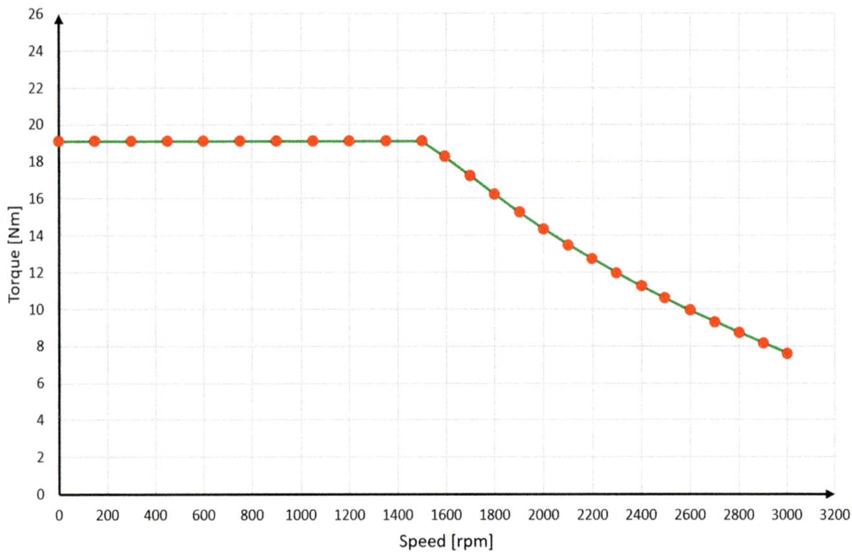

Fig. 13 Optimized design: torque versus speed

Nevertheless, the study can be extended to consider also independent variables for the stator core, resulting in a whole motor optimization.

6 Conclusions

A new strategy to reduce the torque ripple of the synchronous reluctance motor has been presented. The focus was to smooth the torque profile by varying the dimensions of the rotor shape only without any change in the stator core and winding and above all without rotor skewing. For this, a commercial three-phase induction motor has been chosen substituting the squirrel cage rotor with a multiple flux-barriers one. The new rotor has been designed and then refined by an optimization algorithm. A new strategy has been proposed by modelling the particular optimal design problem as mixed integer constrained minimization of a suitable objective function.

References

1. Barcaro, M., Bianchi, N.: Interior PM machines using ferrite to substitute rare-earth surface PM machines. In: 2012 XXth International Conference on Electrical Machines (ICEM), pp. 1339–1345. IEEE (2012)
2. Boldea, I.: Reluctance Synchronous Machines and Drives, vol. 38. Oxford University Press, Oxford (1996)
3. Custódio, A.L., Madeira, J.A., Vaz, A.I.F., Vicente, L.N.: Direct multisearch for multiobjective optimization. SIAM J. Optim. **21**(3), 1109–1140 (2011)
4. Hansen, N.: The CMA evolution strategy: a comparing review. In: Towards a New Evolutionary Computation, pp. 75–102. Springer, Berlin (2006)
5. Hansen, N.: A CMA-ES for mixed-integer nonlinear optimization. Technical report, INRIA (2011)
6. Jahns, T.M., Soong, W.L.: Pulsating torque minimization techniques for permanent magnet AC motor drives-a review. IEEE Trans. Ind. Electron. **43**(2), 321–330 (1996)
7. Liuzzi, G., Lucidi, S., Piccialli, V., Sotgiu, A.: A magnetic resonance device designed via global optimization techniques. Math. Program. **101**(2), 339–364 (2004). https://doi.org/10.1007/s10107-004-0528-5
8. Liuzzi, G., Lucidi, S., Piccialli, V., Villani, M.: Design of induction motors using a mixed-variable approach. Comput. Manag. Sci. **2**(3), 213–228 (2005). https://doi.org/10.1007/s10287-005-0024-2
9. Liuzzi, G., Lucidi, S., Rinaldi, F.: Derivative-free methods for mixed-integer constrained optimization problems. J. Optim. Theory Appl. **164**(3), 933–965 (2015)
10. Liuzzi, G., Lucidi, S., Rinaldi, F.: A derivative-free approach to constrained multiobjective nonsmooth optimization. SIAM J. Optim. **26**(4), 2744–2774 (2016)
11. Liuzzi, G., Lucidi, S., Piccialli, V.: Exploiting derivative-free local searches in DIRECT-type algorithms for global optimization. Comput. Optim. Appl. **65**(2), 449–475 (2016)
12. Moghaddam, R.R., Magnussen, F., Sadarangani, C.: Novel rotor design optimization of synchronous reluctance machine for low torque ripple. In: 2012 XXth International Conference on Electrical Machines (ICEM), pp. 720–724. IEEE (2012)
13. Parasiliti, F., Villani, M.: Synchronous reluctance motors design and performance prediction using an analytical procedure. WIT Trans. Eng. Sci. **41** (2003)
14. Parasiliti, F., Villani, M.: Magnetic analysis of flux barriers synchronous reluctance motors. In: 18th International Conference on Electrical Machines, 2008. ICEM 2008, pp. 1–6. IEEE (2008)
15. Parasiliti, F., Villani, M., Lucidi, S., Rinaldi, F.: Finite-element-based multiobjective design optimization procedure of interior permanent magnet synchronous motors for wide constant-power region operation. IEEE Trans. Ind. Electron. **59**(6), 2503–2514 (2012)
16. Paulavičius, R., Sergeyev, Y.D., Kvasov, D.E., Žilinskas, J.: Globally-biased DISIMPL algorithm for expensive global optimization. J. Global Optim. **59**(2–3), 545–567 (2014)
17. Vagati, A., Pastorelli, M., Francheschini, G., Petrache, S.C.: Design of low-torque-ripple synchronous reluctance motors. IEEE Trans. Ind. Appl. **34**(4), 758–765 (1998). https://doi.org/10.1109/28.703969

A Variant of the Generalized Assignment Problem for Reliable Allocation of Sensor Measurements in a Diagnostic System

Gianmaria De Tommasi, André C. Neto, Antonio Sforza and Claudio Sterle

Abstract Tokamaks are experimental reactors that currently represent the most promising approach to produce electricity by means of the nuclear fusion reactions. In a tokamak a fully ionized gas, called plasma, in which the nuclear reaction occurs, is confined by means of strong magnetic fields. Its performance, physics knowledge and operation safety are significantly affected by several plasma parameters that are controlled by the so called *magnetic control system*. However, these parameters cannot be directly measured, but they are estimated by the so called *magnetic diagnostic*, by combining the information from several sensors. More precisely, different combinations of measurements, which may imply also a different quality of the estimation, can be used to infer the same set of plasma parameters. Each acquisition unit should acquire a set of measurements that allows to reconstruct all the plasma parameters that are needed to operate the machine. In this context, it is fundamental to determine an effective assignment of the sensor measurements to the acquisition units, satisfying capacity constraints and, at the same time, maximizing the overall reliability of the magnetic diagnostic system against possible failures of its components. This problem has been tackled in literature as an original variant of the generalized assignment problem. In this work, we provide a description of the problem itself, as well as a presentation of the related formulations. These formulations are then experienced on real-world test cases in order to show their applicability. Such test cases have been derived from the requirements defined for the JET tokamak, which

---

G. De Tommasi · A. Sforza · C. Sterle (✉)
DIETI, Università di Napoli Federico II, Napoli, Italy
e-mail: claudio.sterle@unina.it

G. De Tommasi
Consorzio CREATE, Napoli, Italy
e-mail: detommas@unina.it

A. Sforza
e-mail: sforza@unina.it

A. C. Neto
Fusion for Energy, Barcelona, Spain
e-mail: Andre.Neto@f4e.europa.eu

C. Sterle
IASI-CNR, Rome, Italy

M. Dell'Amico et al. (eds.), *A View of Operations Research Applications in Italy, 2018*, AIRO Springer Series 2, https://doi.org/10.1007/978-3-030-25842-9_6

is the world's largest tokamak currently in operation. We conclude with a discussion on future research perspectives.

Keywords Generalized assignment · Tokamak · Magnetic diagnostic system

1 Introduction

Nuclear fusion is the reaction in which two or more atomic nuclei fuse to form one or more heavier different atomic nuclei and subatomic particles. The fusion is made using light nuclei, specifically hydrogen and its isotopes (i.e. deuterium and tritium), since their limited dimensions and the small number of protons allow the nuclear force to more easily overcome the repulsion. The difference in mass between the reactants and products provides the release of energy. Nuclear fusion presents opposite characteristics with respect to nuclear fission and it has several theoretical advantages: ample fuel supplies, since hydrogen is in practice unlimited; reduced radioactivity, little nuclear risk and waste, since there is no chemical production from combustion. However, nowadays, controlled fusion is still a challenge, given the difficulty to produce it in a practical and economical manner.

The *tokamak* is an experimental machine conceived to exploit energy released by the fusion to produce steam and electricity by turbines and generators. The functioning principle is the following: a column of fully ionised gas (*plasma*), where the reaction takes place, is confined by means of external magnetic fields, generated by a set of *toroidal* and *poloidal* coils distributed around the vacuum vessel, in a torus shape and heated by external sources (see Fig. 1). The aim is to achieve the condition where the *gain* Q is greater than one. This gain expresses the ratio between the energy released by the fusion process and the energy used to sustain the fusion in a steady state. The so called break-even condition is achieved when $Q = 1$, while when $Q > 1$, the energy provided by the fusion reaction may be captured and used for self-heating and theoretically the *ignition* condition can be reached. The Joint European Torus (JET), in Culham (Oxfordshire, UK), which was built in the early 1980s, is the world's largest operational tokamak, and it was explicitly designed to study the plasma behaviour during the fusion process. During its operation it reached $Q \approx 0.67$. The next step in fusion experimental reactors is ITER, which is under construction in Cadarache (France), which will be the world's largest tokamak, designed to reach $Q = 10$.

The performance of a tokamak, its physics knowledge and operation safety are significantly affected by several plasma parameters, e.g.: the shape, the vertical position and velocity, and the plasma and toroidal current. The technical explanation of these parameters is out of the scope of this work, and the interested reader may refer to [5]. Here, we just point out that, for effective and safe operations, it is fundamental to opportunely control them. However, these parameters cannot be directly measured, but they can be estimated combining the information provided by several sensors measuring the magnetic fields and fluxes on the boundary of the reactor. These sen-

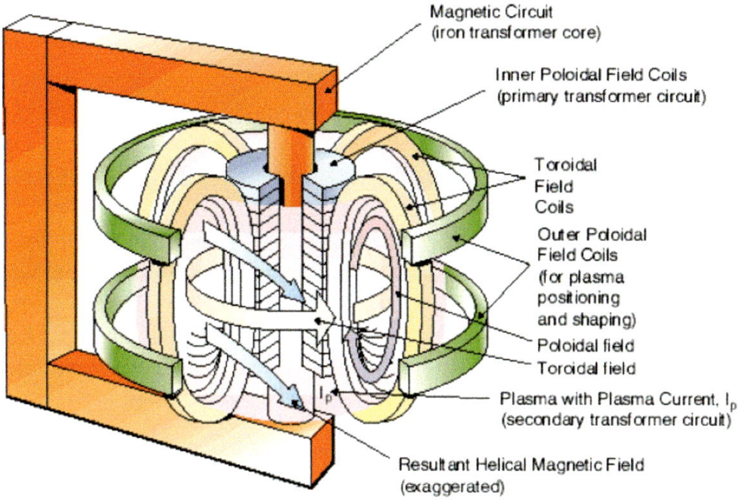

Fig. 1 Schematic representation of a tokamak

sors constitute one of the key component of the so called *magnetic diagnostic system* and they are acquired a processed by the data acquisition units (also called *fast plant system units*, FPSU), to infer the required parameters [8, 10, 13]. In a tokamak, thousand of sensors are spread around the sectors of the reactor and they are generally grouped according to their type and/or location. Under the ideal assumptions of axisymmetry and absence of noise, the measurements from only one sector should be sufficient to reconstruct all the needed parameters with the required accuracy. However, each group is usually replicated in order to increase the redundancy of the system and to get more than one estimation for each parameter of interest. Moreover, the redundancy is still increased by the fact that diverse measurement combinations, differing in the estimation quality, can be used to infer the same parameter. These redundancies are fundamental for the overall system reliability, expressed in terms of number of times each parameter can be estimated in case of failures of the data acquisition components of the magnetic diagnostic system.

In case of faults that may lead to a shutdown of FPSUs, it follows that, the local parameter estimation is possible if all the needed information, provided by the sensors, are stored in the same FPSU. Hence, given a data acquisition architecture, composed of several FPSUs, it is fundamental to determine an allocation of the sensor measurements in the available slots, satisfying the capacity constraints and, at the same time, maximizing the overall reliability of the magnetic diagnostic system against possible failures of its components. This problem has been defined in literature as the optimal sensor measurement allocation problem and it has been modeled as an original variant of the *Generalized Assignment Problem* (GAP) in [3]. GAP is a well-known NP-hard combinatorial optimization problem, and the interested reader is addressed to [1, 4, 9] for more details. Here, we just recall its more general definition. Given *m bins* of fixed capacity (K_m) and *n objects*, characterized by weights and profits depending on the pairs object-bin, the GAP consists of determining the maximum profit assignment of objects to bins, such that each object is assigned to exactly one bin, and capacity constraints are satisfied [6]. In the proposed variant of the GAP, the objects to be allocated are the sensor measurements and the bins are the FPSUs. The profit to be *maximized* is given by the reliability of the system, expressed in terms of number of estimations of each parameter.

This work is aimed at presenting the formulations proposed in literature for the optimal sensor allocation problem and showing their applicability to real-world test cases [2, 3, 11]. The paper is structured as follows: Sect. 2 is devoted to the problem definition; Sects. 3 and 4 report the mathematical programming formulations proposed in literature, together with their applications to the real-world test cases; finally, some conclusive remarks and work perspectives are given in Sect. 5.

2 Problem Definition

The magnetic diagnostic system of a modern tokamak device, such as the ITER tokamak, exploits a significant number of magnetic sensors, whose measurements are combined to estimate the relevant plasma control parameters [12]. There are currently ~2000 sensors, divided into groups sharing properties, such as, the type of technology or the toroidal section. Measurements of the sensors belonging to the same group are generally stored in the same FPSU. Hence, in the following, without loss of generality, we will not refer to a single sensor, but to sensor groups.

The basic architecture of the system consists of six different FPSUs composed of several slots [7]. Each FPSU can continue to independently operate if one of the other fails, as well as a slot of a FPSU can operate in case of failures of the others. Given this architecture, thanks to the above explained system redundancies, the same plasma parameter can be computed using different combinations of sensor group measurements stored in the same FPSU. Moreover, the same sensor group measurement can concur to compute different parameters. Hence, the assignment of a group to a FPSU represents a strategic choice.

In order to better explain the effect of this choice, let us consider the following small example with five sensor groups: $\mathscr{S} = \{A, B, C, D, E\}$. Two groups G_1 and G_2 are *redundant* if the sensor measurements of G_1 can be used in place of the ones of G_2 and viceversa to estimate the control parameters. On this basis, let us consider the following redundant groups: $a = \{A \cup B \cup C\}$ and $b = \{D \cup E\}$, i.e. in the estimation of a parameter the sensor groups in A can be replaced by the sensors in B or in C, while the group D can be replaced by E.

Let f_1 and f_2 be two plasma parameters to be estimated. In particular f_1 can be expressed as $f(\mathscr{S})$, i.e. it can be computed using a single group of $\mathscr{S}$. Hence, the following five alternatives are available: $f_{11}(A)$, $f_{12} = f(B)$, $f_{13} = f(C)$, $f_{14} = f(D)$ and $f_{15} = f(E)$. Instead, f_2 can be expressed as $f(a, b)$, i.e. it can be computed using a redundant group of subset a and a redundant group of subset b. Hence, the following six possible alternatives are available: $f_{21}(A, D)$, $f_{22} = f(A, E)$, $f_{23} = f(B, D)$, $f_{24} = f(B, E)$, $f_{25} = f(C, D)$, and $f_{26} = f(C, E)$. Given the previous alternatives, we can also define the so called *redundancy level* of a parameter f with g redundant groups as:

$$\rho_f = \min\left(\gamma_1, \ldots, \gamma_g\right),$$

where γ_i is the cardinality of the generic redundant set i used to estimate a parameter f. For the considered example, with respect to f_1, we have $\gamma_{\mathscr{S}} = 5$ and hence $\rho_{f_1} = 5$. Instead, for f_2, we have $\gamma_a = 3$, $\gamma_b = 2$, hence $\rho_{f_2} = 2$.

Each alternative of a plasma parameter is associated with a different *profit*, i.e. a numerical value which expresses the importance of the parameter and the quality of its estimation. Let us assume the following values: 100, 90, 20, 50 and 40 for f_1; 50, 45, 45, 50, 30, 30 for f_2. Let us consider the two parameters independently and let us also recall that the reliability of the system depends on the number of alternatives that can be used to estimate a parameter in case of failure of a FPSU or part of its slots. In order to maximize the reliability of the system with respect to f_1, we should use five different FPSUs and assign a group to each of them. Considering instead f_2, we should use two FPSUs and we would assign groups A and D to the first and groups B and E to the second. Group C could be either assigned to one of the already used FPSUs or to a different one, depending on the capacity constraints. This allows to estimate parameter f_2 with the two alternatives characterized by the highest profit values.

It is easy to understand, that both solutions present some drawbacks. The first allows to achieve the full redundancy of parameter f_1 characterized by the highest profit values, but does not allow to compute any of the alternatives of parameter f_2. On the contrary, second solution allows to cover both parameters, guarantees the full redundancy of f_2, but it looses something in terms of overall profit and f_1 redundancy.

A good allocation of the sensor groups in the FPSUs should guarantee the following requirements in order to be effective and reliable [2, 11]:

- **R0** all the sensor group measurements are stored in one of the available FPSUs;
- **R1** the maximum number of plasma control parameters is covered;
- **R2** the alternatives of a parameter have to be allocated in the highest number of different FPSUs (*first level redundancy*);
- **R3** the alternatives of a parameter that can be computed in the same FPSU should be maximized (*second level redundancy*).

The order of these requirements is a priority order and in general it is not possible to meet simultaneously all the them, unless all the parameters require distinct and disjoint sensor groups to be estimated or it is possible to have replicas of the same measurements in different FPSUs and capacity requirements are satisfied. Therefore a good trade-off must be found and the formulations presented in the following sections are aimed at achieving this task.

3 Problem Setting and Basic GAP Based Formulation

The following problem setting is required to formulate the problem by an integer linear programming formulation:

- $\mathcal{S}$ is the set of sensor groups whose cardinality is S
- $\mathcal{F}$ is the set of parameters to be estimated whose cardinality is F
- N_i is the set of possible alternatives for the ith parameter, with $i = 1, \ldots, F$. Hence, $N = \sum_{i=1}^{F} N_i$ is the number of all possible alternatives.
- $A \in \{0, 1\}^{N \times S}$ is a binary matrix indicating which sensor groups are involved in the estimation of the different parameter alternatives. The generic element a_{js} is equal to 1 if the sth group is used to compute the jth alternative, 0 otherwise.

Moreover, the following parameters are used to specify the cost function to be maximized:

- w_s is the *weight* of the sth group of sensors, i.e. the number of sensors composing the sth group, $s = 1, \ldots, S$.
- K is the capacity of a FPSU, i.e. the maximum number of sensors that can be allocated in one FPSU.
- R is the number of available FPSUs.
- μ_i is the minimum between R and the full redundancy level ρ_i of the ith parameter, $i = 1, \ldots, F$.
- $p_{i,j}$ is the *profit* of the ith parameter when the jth alternative can be estimated by sensor groups assigned to a single FPSU.
- λ is a *penalty* factor which is used to negatively affects the objective function in case more alternatives of the same parameter are allocated on the same FPSU or a number of alternatives lower than μ_i can be estimated.

Finally, the following decision variables are used:

- y_s^r, binary variable equal to 1 if the sth sensor group, $s = 1, \ldots, S$, is allocated in rth FPSU, $r = 1, \ldots, R$; 0 otherwise.
- z_i^r, binary variable equal to 1 if at least one alternative of the ith parameter, $i = 1, \ldots, F$, can be estimated with the sensor measurements stored in rth FPSU, $r = 1, \ldots, R$; 0 otherwise.
- x_{ij}^r binary variable equal to 1 if the jth alternative, $j = 1, \ldots, N_i$, of the ith parameter, $i = 1, \ldots, F$, can be computed with the sensor measurements in the rth FPSU, $r = 1, \ldots, R$; 0 otherwise.
- θ_i, is a general integer variable counting the number of FPSUs used to store the allocated alternatives of the ith parameter, $i = 1, \ldots, F$.

On this basis, the problem can be formulated by the following ILP model [3]:

$$(BASIC) \qquad \max \left[\sum_{i=1}^{F} \sum_{j=1}^{N_i} \sum_{r=1}^{R} p_{i,j} x_{ij}^r - \lambda \sum_{i=1}^{F} (\mu_i - \theta_i) \right] \tag{1}$$

subject to:

$$\sum_{r=1}^{R} y_s^r = 1, \qquad \forall s \in \mathscr{S} \tag{2}$$

$$\sum_{s=1}^{S} w_s y_s^r \leq K, \qquad \forall r \in R \tag{3}$$

$$\sum_{s=1}^{S} a_{js} y_s^r \geq \text{SUM}_j \cdot x_{ij}^r, \qquad \forall i \in F, j \in N_i, r \in R \tag{4}$$

$$\sum_{r=1}^{R} x_{ij}^r \leq 1, \qquad i \in F, j \in N_i \tag{5}$$

$$\sum_{j=1}^{N_i} x_{ij}^r \geq z_i^r, \qquad i \in F, r \in R \tag{6}$$

$$\sum_{r=1}^{R} z_i^r = \theta_i, \qquad i \in F \tag{7}$$

$$y_s^r \in \{0, 1\}, \qquad s \in \mathscr{S}, r \in R \tag{8}$$

$$z_i^r \in \{0, 1\}, \qquad i \in F, r \in R \tag{9}$$

$$x_{ij}^r \in \{0, 1\}, \qquad i \in F, j \in N_i, r \in R \tag{10}$$

$$\theta_i \in N, \qquad i \in F \tag{11}$$

where $\text{SUM}_j = \sum_{s=1}^{S} a_{js}$.

The objective function (1) has two terms trying to achieve the trade-off between the allocation of the maximum number of the overall available alternatives, weighted by their profit (referred as *Profit* in what follows), and the distribution of these alternatives in different FPSUs (in the following referred as *Redundancy*). In particular: the first term maximizes the profit associated to the parameter alternatives, whereas the second pushes towards the spreading of these alternatives in different FPSUs. Indeed, it contributes negatively in the objective function value if $\theta_i < \mu_i$ or if $\theta_i = 0$ and it provides no contribution just in case $\theta_i = \mu_i$ for all the parameters. Hence, the lowest is this term, the better is for the overall objective function value. We highlight that although the constant term $-\lambda \sum_{i=1}^{F} \mu_i$ could be removed from the model, we used it in (1) to better highlight the effect of the penalty value. Furthermore:

- Constraints (2) and (3) are classical *GAP* constraints. In particular, Constraints (2) impose that no replicas of the same sensor group can be created. To allow solutions with replicas, it is sufficient to impose that the left-hand side has to be higher than or equal to the right-hand side.
- Constraints (4) impose that if all the sensor groups contributing to the jth alternative of the ith parameter are allocated in the same rth FPSU, then the corresponding alternative can be actually estimated.

- Constraints (5) impose that the jth alternative of the ith parameter can be estimated by at most one FPSU.
- Constraints (6) are consistency constraints between the optimization variables x_{ij}^r and z_i^r.
- Constraints (7) compute the θ_i values for each parameter.
- Constraints (8), (9), (10), and (11) define the domain of the problem.

In the following section we provide an application of the basic model to a test-case, in order to better highlight its main drawbacks and to better understand the improvements proposed in Sect. 4.

3.1 Application of the Basic Model to a Real-World Test Case

The basic model has been tested on an instance considering 1087 sensors grouped in 70 groups used to estimate 13 plasma parameters with an overall number of alternative equal to 49. The sensors have to be allocated in the 6 available FPSUs, each of them with capacity equal to 336 slots. We recall that each slot can store the information of one sensor belonging to a given group. This instance has been generated by deriving the control parameters and the correspondent alternatives from the ones currently used at JET, and by considering the ITER magnetic diagnostic system architecture as reference (more details on the architecture proposed for ITER can be found in [3, 7]).

The model has been solved with a computation time of few minutes by the optimization software FICO(TM) Xpress-MP 8.0 and run on an Intel(R) Core(TM) i7-8700k, 3.70 GHz, 16.00 GB of RAM, using default setting [3]. The main input information of the instance under investigation and related results are shown in Figs. 2a, b, respectively. In particular, Fig. 2b reports for different λ values (1, 5 and 10) the first level redundancy of each parameter, i.e. the number of alternatives assigned to different FPSUs that can be computed with the resulting sensor allocation. Figure 3 reports for all the values of λ within the range [0, 100] with step 0.1, the obtained solutions in terms of *Profit* and *Redundancy*, given by (1), and in terms of number of estimated functions and computed alternatives.

These results show that even if several solutions provide the same number of estimated functions and computed alternatives, they are not equivalent in terms of redundancy and profit values. This is due to the fact that, as expected, the values of λ play a crucial role. Indeed, lower values of the penalty factor provide solutions covering a higher number of parameters, whereas high values of the penalty factor push towards solution where few parameters characterized by high full redundancy level values are covered.

It is important to note that, before the introduction of the problem formulation presented in Sect. 3, the designers of the ITER magnetic diagnostic data acquisition system were using a *rule-based* allocation algorithm. In particular, the rules of such an algorithm were derived from the past experience, and the correspondent

(a)

ID	Processed signal	ρ_i	N_i	$\mu_i = \min(R, \rho_i)$
f_1	BETA-DIA	2	4	2
f_2	FISHBONE-AMP-PHASE	1	1	1
f_3	IHALO-IP	1	1	1
f_4	IP-CIS	6	6	6
f_5	IP-HF	1	1	1
f_6	IVV	2	2	2
f_7	SHAFRANOV-INT	2	3	2
f_8	VLOOP-CFLI	8	8	6
f_9	VLOOP-CFLO	4	4	4
f_{10}	VLOOP-FL	6	6	6
f_{11}	VLOOP-MHD	8	8	6
f_{12}	dz_dt-IP	2	2	2
f_{13}	dz_dt-IP-n0	6	6	6

(b)

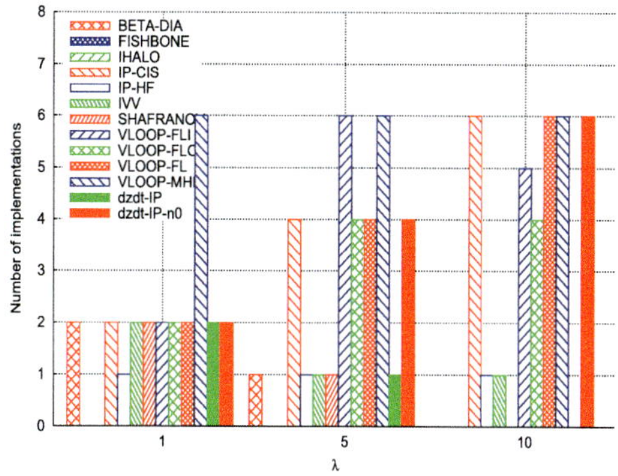

Fig. 2 **a** Instance input information. **b** Results of the basic model: first level redundancy for three different λ values

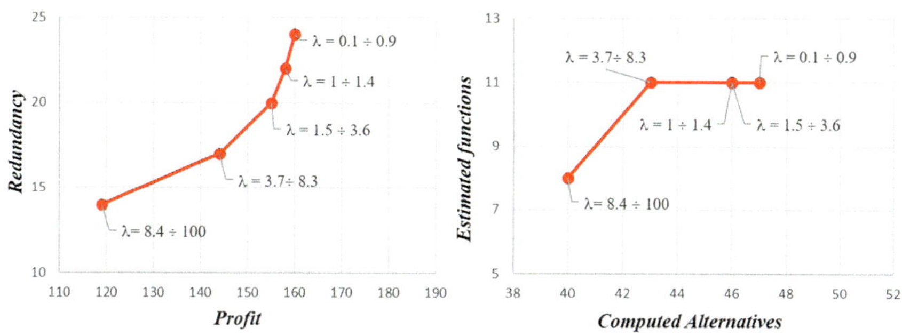

Fig. 3 Basic model solutions for different values of the λ parameter

parameters were tuned with according to a *trial-and-error* approach, aiming at improving the reliability of the overall system. In practice, it turned out that the tuning of the algorithms parameters was not straightforward, since they were many and their impact on the overall system performance was not clear, since it was not formalized. Hence, the availability of a *optimization-based* approach was immediately successful, since it allows to clearly define the overall objective, as well as to derive some insights on the tuning of the profit values for each parameter to be controlled. Moreover, the availability of a formal description of the problem to be solved permitted to easily spot out some flaws of the approach described in Sect. 3, leading to the proposal of the improved GAP formulation presented in the next section.

4 Improved GAP Based Formulation

The formulation presented in the previous section was the first one tackling the problem under investigation. Its objective function is composed of two terms related to two optimization criteria which are evaluated using different units of measure. This would naturally define a multi-objective optimization problem, which, in the *BASIC* model, is solved considering a linear combination of the optimization criteria. The usage of this straight solving method has a drawback. Indeed, the model can return good solution in terms of requirements **R0**, **R2** and **R3**, but it does not well respond to requirement **R1**. In other words, the estimation of a further alternative of a high profit parameter is preferred to the estimation of an alternative of another yet uncovered parameter. Moreover, even if both redundancy level are taken into account, the priority among them is not guaranteed for the same reason provided for **R1**. To avoid this occurrence, a more appropriate multi-objective optimization method would be required. However, we did not investigate this issues because we developed an improved ILP formulation which overcomes the *BASIC* model drawback [2, 11]. To this aim we have to redefine the profit of an alternative as follows:

- $p^1_{i,j}$, is the profit achieved when the jth alternative of the ith parameter is *the first one* to be estimated;
- $p^2_{i,j}$, is the profit achieved when the jth alternative of the ith parameter is *not the first one* to be estimated.

These two profit values are strictly connected with the concepts of first and second redundancy levels defined above. This aspect will be better clarified when explaining the improved formulation. Moreover, the following decision variables are introduced:

- $x^{1,r}_{ij}$, binary variable equal to 1 if the jth alternative of the ith parameter is considered as *the first one* to be estimated by sensor groups measurements in the rth FPSU; 0 otherwise.
- $x^{2,r}_{ij}$, binary variable equal to 1 f the jth alternative of the ith parameter is considered as *not the first one* to be estimated by sensor group measurements in the rth FPSU; 0 otherwise.

Given the above definitions, the new formulation of the problem is the following [2, 11]:

$$(IMP) \qquad \max \left[\sum_{i=1}^{F} \sum_{j=1}^{N_i} \sum_{r=1}^{R} p^1_{i,j} x^{1,r}_{ij} + \sum_{i=1}^{F} \sum_{j=1}^{N_i} \sum_{r=1}^{R} p^2_{i,j} x^{2,r}_{ij} \right] \tag{12}$$

subject to:

(2), (3), (4), (5)

$$\sum_{j=1}^{N_i} x^{1,r}_{ij} \le 1, \qquad i \in F, r \in R \tag{13}$$

$$x^r_{ij} \ge x^{1,r}_{ij} + x^{2,r}_{ij}, \qquad i \in F, j \in N_i, r \in R \tag{14}$$

The new objective function maximizes the first and second level redundancy (**R2** and **R3**), whereas the requirements **R0** and **R1** are guaranteed by the constraints. The values of $p^1_{i,j}$ and $p^2_{i,j}$ have to be opportunely defined in order to avoid solutions where just few alternatives of high profit parameters are estimated and no alternatives of low profit parameters can be estimated. To this aim, all the $p^2_{i,j}$ values have to be lower or equal to the minimum value of $p^1_{i,j}$, $\forall i \in F$ and $\forall j \in N$. In this way, a first alternative of an uncovered function will be always preferred with respect to a second alternative of an already covered function. At the same time, if a sensor group cannot be exploited for the estimation of a first alternative of an uncovered parameter, but it can be used to replace another group in the estimation of an alternative of an already covered parameter, then it is allocated on the same FPSU. Constraints (13) impose that only one first alternative of a parameter can be allocate on each FPSU. Constraints (14) impose that if an alternative is allocated on a FPSU, then either it is the first alternative or it is the second one. The objective function, maximizing the overall reliability of the system, will select as first alternative the one associated with the highest profit, and will consider all the others as second alternatives.

4.1 Comparison Between the Proposed Formulations on a Real-World Instance

In this section we show the different results obtained by the two proposed formulations (BASIC and IMP). We consider an instance, different from the one of previous test case, since it better shows the advantages of the improved formulation with respect to the basic one. The new instance, also derived from the JET data available, considers 2000 sensors, grouped in 94 groups to be allocated in 6 FPSUs each of the with capacity equal to 336 slots. There are 32 parameters to be estimated and a total number of 69 alternatives. The results of the two formulations in terms of number of alternatives covered for each parameter are shown in Fig. 4. For the basic model, we

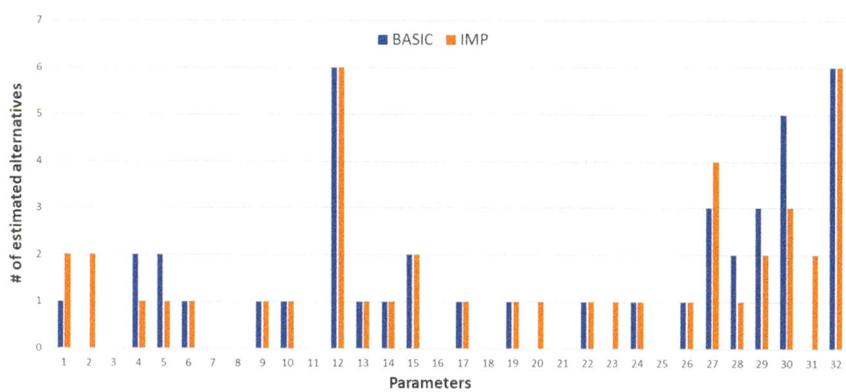

Fig. 4 Comparison between the proposed formulations

considered the best solution obtained with $\lambda = 1$. We observe that even if the number of alternatives estimated by the two solutions is quite similar, 42 for BASIC and 44 for IMP, there is a significant difference in the number of estimated parameters, 20 for BASIC and 24 for IMP. This is mainly due to the fact that in BASIC model, although we used low values of the penalty factor, the parameters with higher profit values significantly affect the solution.

5 Conclusions

In this work we presented two integer linear programming formulations proposed for the optimal allocation of the sensor measurements in a tokamak magnetic diagnostic system. The two proposed formulations have been tested on two real-world test cases providing interesting results and showing their applicability. It is important to remark that, before adopting the proposed *optimization-based* approach for signal allocation, the system designers were using a *trial-and-error* rules-based algorithm, whose tuning was very difficult. Moreover, since the objective was not formally defined, with the previous approach it was not possible to measure the *optimality* of the obtained solution. The system experts considered the proposed methodologies as useful decision support tools for two main reasons: on one side, they allow to better tune the profit values of the parameters, as well as to better weight the relative importance among them. On the other side, the proposed approach allows them to identify system criticalities in terms of uncovered parameters and redundancy requirements. This provides also with hints about the number and the dimension (in terms of slots) of FPSUs to be used in the diagnostic system in order to achieve the required system reliability. Future research activity could consider the introduction of failure probabilities for each system component and, consequently, the development of ad-hoc robust optimization techniques. Finally, it is worth to notice that the proposed approach can be easily extended to all the diagnostic system of a complex infrastructure wherein the quantities of interest can be obtained by processing the measurements acquired by different spatially distributed sensors.

References

1. Cattryssea, D., Wassenhove, L.V.: A survey of algorithms for the generalized assignment problem. Eur. J. Oper. Res. **60**(3), 260–272 (1992)
2. De Tommasi, G., Neto, A., Pironti, A., Sterle, C.: Optimal allocation of the diagnostic signals for the iter magnetic control system. In: 2015 IEEE Conference on Control Applications (CCA), pp. 1296–1302. Sidney, Australia (2015)
3. De Tommasi, G., Neto, A., Sterle, C.: PIMPA: a tool for optimal measurement probes allocation. IEEE Trans. Plasma Sci. **42**(4), 976–983 (2014)
4. Hopcroft, J., Motwani, R., Ullman, J.: Introduction to Automata Theory, Languages, 3rd edn. Prentice Hall, Upper Saddle River (2007)

5. Hutchinson, I.: Principles of Plasma Diagnostics. Cambridge University Press, Cambridge (2005)
6. Karlof, J.: Integer Programming: Theory and Practice. Taylor & Francis, Broken Sound Parkway, NW (2006)
7. Neto, A., et al.: Conceptual architecture of the plant system controller for the magnetics diagnostic of the ITER tokamak. Fusion Eng. Des. **96**, 887–890 (2015)
8. Pironti, A., Walker, M.: Fusion, tokamaks, and plasma control: an introduction and tutorial. IEEE Control Sys. Mag. **25**(5), 30–43 (2005)
9. Posta, M., Ferland, J., Michelon, P.: An exact method with variable fixing for solving the generalized assignment problem. Comput. Optim. Appl. **52**(3), 629–644 (2012)
10. Quercia, A., et al.: Performance analysis of Rogowski coils and the measurement of the total toroidal current in the ITER machine. Nucl. Fusion **57**(12), 126049 (2017)
11. Sterle, C., Neto, A., De Tommasi, G.: An improved model for the optimal measurement probes allocation tool. Fusion Eng. Des. **96**, 970–973 (2015)
12. Vergara Fernández, A., et al.: Modeling tools for the ITER central interlock system. Fusion Eng. Des. **86**(6–8), 1137–1140 (2011)
13. Wesson, J.: Tokamaks. Oxford University Press, Oxford (2004)

Production Planning and Scheduling

Forecasting Methods and Optimization Models for the Inventory Management of Perishable Products: The Case of "La Centrale del Latte di Vicenza SpA"

Luca Bertazzi and Francesca Maggioni

Abstract This paper is a report on a joint project by the Department of Management, Economics and Quantitative Methods of the University of Bergamo (in collaboration with the Department of Economics and Management of the University of Brescia) and the Italian company "La Centrale del Latte di Vicenza SpA" producing different types of milk, yogurt, vegetable drinks, cream, butter, cheese, eggs and vegetables. The aim of the project was to provide forecasting methods and optimization models, to improve the demand forecasts of perishable products and to better manage inventory levels in a *Material Requirements Planning* (MRP) setting.

Keywords Perishable products · Forecasting methods · Material Requirements Planning (MRP) · Optimization models

1 Introduction

Demand forecasting and optimal allocation of resources are critical processes for any company. Managers typically have to take complex decisions in short time. They have to analyze larger and larger amount of data and qualitative information, often in an uncertain setting. Quantitative methods (like statistical methods, Operations Research models and algorithms, ...) can be adopted to support managers in taking these decisions (see [4]).

The purpose of this article is to describe the forecasting methods and the *Material Requirements Planning* (MRP) optimization models we provided to support an Italian company, "La Centrale del Latte di Vicenza SpA", interested in having better demand forecasts and a better inventory management of perishable products. We

---

L. Bertazzi
Department of Economics and Management, University of Brescia, Brescia, Italy
e-mail: luca.bertazzi@unibs.it

F. Maggioni (✉)
Department of Management, Economics and Quantitative Methods, University of Bergamo, Bergamo, Italy
e-mail: francesca.maggioni@unibg.it

M. Dell'Amico et al. (eds.), *A View of Operations Research Applications in Italy, 2018*, AIRO Springer Series 2, https://doi.org/10.1007/978-3-030-25842-9_7

supported the Company in identifying the best forecasting methods between the *Locally Weighed Regression Estimator* (LOESS) and the *Seasonal Holt-Winters* (HW). Then, we formulated optimization models replicating the MRP used in the Company. An optimal solution of these models allowed us to evaluate the MRP embedded in the information system used by the Company and to reduce the inventory levels of raw materials, packaging materials, components and final products. We applied these methods and models to a large set of real instances provided by the Company.

The structure of the paper is as follows. In Sect. 2 we give a description of the real problem and of the joint project. In Sect. 3 we describe the forecasting methods we suggested to the Company. In Sect. 4 we describe the simplest optimization model we formulated to evaluate and improve the MRP used by the Company. In Sect. 5 we provide examples of the application of these methods and models to real instances provided by the Company. Conclusions are drawn in Sect. 6.

2 The Real Problem and the Joint Project

"La Centrale del Latte di Vicenza SpA" produces and packages different formats and types of milk, yogurt, vegetable drinks, cream, butter, cheese, eggs and vegetables. These products have different shelf life, ranging for example from 5 days for fresh milk, to 50 days for yogurt and to 120 days for UHT milk. All articles follow a pull production, with exception of UHT milk for which a push system is used. The Company buys from suppliers raw materials—jams and lactic ferments—as well as materials for packaging. Some of these products are perishable and, therefore, if they are not used before their expiry date, additional costs are paid by the Company. Reliability of the suppliers, delivery times, lot sizes have to be taken into account in the production management.

Before this project, the Company used the MRP embedded in the information system as a black box to make requirements planning, without applying any quantitative method to forecast the demand of the final products, but just monitoring the sales trend on a daily basis. For production on behalf of third parties, in the presence of a high demand uncertainty, the Company considered monthly sales budgets and divided it into a weekly time frame.

The aim of the Company in this project was, on one side, to start using quantitative methods to forecast the demand, especially for products on behalf of third parties, identifying possible trends and seasonality. On the other side, the aim was to start using optimization models to have an optimal requirements planning, having in this way a tool to evaluate the MRP used by the Company and to improve it by reducing the inventory levels. Note that these optimization models can use the demand forecasts of the final products, provided in the first part of the project, as an input data.

3 Forecasting Methods

Concerning the demand forecast of the final products, we suggested the following methods to the Company: *Locally Weighed Regression Estimator* (LOESS) and *Seasonal Holt-Winters* (HW). Here we briefly recall them.

LOESS is a modern modeling methods originally proposed in [1], designed to address situations in which the classical procedures do not perform well. It fits simple models to localized subsets of the data to build up a function that describes the deterministic part of the variation in the data, point by point. The main attraction of this method is that it does not require to specify a global function to fit to the data, but only to fit segments of the data. At each point in the data set a low-degree polynomial is fit to a subset of the data, with explanatory variable values near the point whose response is being estimated. The polynomial is fit using weighted least squares, giving more weight to points near the point whose response is being estimated and less weight to points further away. The value of the regression function for the point is then obtained by evaluating the local polynomial using the explanatory variable values for that data point. The LOESS fit is complete after regression function values have been computed for each of the n data points. Many of the details of this method, such as the degree of the polynomial model and the weights, are flexible. The subsets of data used for each weighted least squares fit in LOESS are determined by a nearest neighbors algorithm. A user-specified input to the procedure q called smoothing parameter determines how much of the data is used to fit each local polynomial. q is a number between $(d + 1)/n$ and 1, with d denoting the degree of the local polynomial. The subset of data used in each weighted least squares fit is comprised of the nq (rounded to the next largest integer) points whose explanatory variables values are closest to the point at which the response is being estimated. Useful values of the smoothing parameter typically lie in the range 0.25–0.5 for most LOESS applications.

The local polynomials fit to each subset of the data are almost always of first or second degree. Using a zero degree polynomial turns LOESS into a weighted moving average (MA).

The traditional weight function used for LOESS is the tri-cube weight function,

$$W(u) = \begin{cases} (1 - u^3)^3 & 0 \leq u < 1 \\ 0 & otherwise \end{cases} \qquad (1)$$

The weight for a specific point in any localized subset of data is obtained by evaluating the weight function at the distance between that point and the point of estimation, after scaling the distance so that the maximum absolute distance over all of the points in the subset of data is exactly one.

HW is a method that allows us to take into account trend and seasonality in the data (see [2, 5]). It comprises the forecast equation and three smoothing equations, one for the level e, one for the trend T, and one for the seasonal component S. Here we provide an additive type model which takes into account both the trend and the seasonality (it is possible to consider also a multiplicative type model). The forecast model for the period $i+\theta$ (with $\theta = 1, 2, 3\ldots$) is:

$$\widehat{Y}_{i+\theta} = \widehat{e}_i + \widehat{T}_i\theta + \widehat{S}_{i+\theta-s} \tag{2}$$

where $\widehat{S}$ represents the estimate of the seasonality, $\widehat{e}$ the estimate of the level and s is the seasonality period. The exponential smoothing models to be used to estimate the average level $\widehat{e}_i$, trend $\widehat{T}_i$ and seasonality $\widehat{S}_i$ at time i are:

$$\widehat{e}_i = \delta_1(Y_i - \widehat{S}_{i-s}) + (1-\delta_1)(\widehat{e}_{i-1} + \widehat{T}_{i-1}) \tag{3}$$

$$\widehat{T}_i = \delta_2(\widehat{e}_i - \widehat{e}_{i-1}) + (1-\delta_2)\widehat{T}_{i-1} \tag{4}$$

$$\widehat{S}_i = \delta_3(Y_i - \widehat{e}_i) + (1-\delta_3)\widehat{S}_{i-s} \tag{5}$$

where δ_1, δ_2 e δ_3 are the corresponding smoothing parameters respectively for the level e_i, the trend T_i and the seasonality S_i.

The three smoothing parameters δ_1, δ_2 e δ_3 lie between 0 and 1 and their optimal values δ_1^*, δ_2^* and δ_3^* can be found by minimizing the sum of the squares of the prediction errors.

4 An Optimized Material Requirements Planning (MRP)

In the second part of the project we provided optimization models replicating the *Material Requirements Planning* (MRP) used by the Company, with the goal to evaluate the MRP embedded in the information system used by the Company and to find better solutions in terms of inventory levels.

The MRP is a requirements planning tool used when a final product is produced by assembling material raws and components (each referred to as *Stock Keeping Unit*—SKU) in successive steps, on the basis of a given *Bill of Materials* (BOM). Starting from the demand forecast of the final product, the MRP allows to the managers to decide when to buy or to produce each SKU and in which quantity. The aim is to buy or to produce each SKU as late as possible, without having stock–out.

We built a user-friendly application in Excel, by using the programming language *Visual Basic for Applications* (VBA). This application, given the data provided by the information system of the Company, is able to compute all advanced data needed for the model, to run an optimization solver (*Solver studio* [7] in our case) and to import the optimal solution in Excel (see Fig. 1). The availability of the optimal solution in Excel is useful, for example, to compare the solution used by the Company with the optimal solution.

The first step was to define the set of data, available in the information system of the Company, to use in the model. The identified the following data:

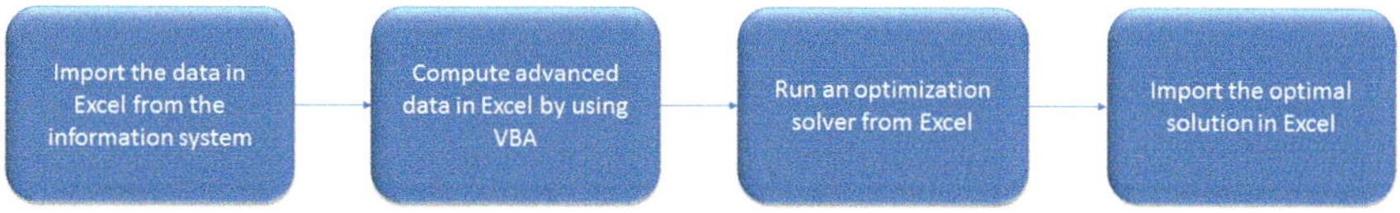

Fig. 1 An optimized MRP

1. P: Number of SKU (SKU 1 is the final product)
2. T: Number of time buckets (i.e. the planning horizon)
3. D_t: Demand of the final product at time t
4. C_t: Planned delivery of the final product at time t (from previous planning)
5. S_t: Quantity of the final product to produce at time t from previous planning
6. π_t: Equal to 1 if the production of the final product is feasible at time t and 0 otherwise
7. LS_i: Lot size of the SKU i
8. $\mathscr{F}$: Set of SKU i for which the quantity produced or purchased is a multiple of the lot size LS_i
9. $\mathscr{S}$: Set of SKU i for which the quantity produced or purchased is greater or equal than the lot size LS_i
10. LT_i: Lead time of the SKU i
11. $\bar{I}_i$: Starting inventory level of the SKU i
12. R_{ij}: Number of units of SKU i to produce 1 unit of SKU j (from the *Bill of Materials*—BOM)
13. O_{it}: Number of units of SKU i made available at time t thanks to previously placed orders
14. IA_{it}: Number of units of SKU i used at time t to produce other final products
15. FA_{it}: Number of units of SKU i scheduled at time t to produce other final products
16. F_i: Minimal number of units of SKU i to produce whenever the SKU i is produced
17. M: sufficiently large number.

The second step was to formulate an optimization model replicating the MRP. We first introduced the following decision variables:

1. x_{it}: non–negative continuous variable representing the number of units of SKU i produced or purchased at time t
2. $\hat{x}_{it}$: non–negative continuous variable representing the net number of units of SKU i produced or purchased at time t (for the final product this is equal to $x_{it} - S_t$, while it is equal to x_{it} for the other SKUs)
3. I_{it}: non–negative continuous variable representing the inventory level of SKU i at the end of time bucket t
4. z_{it}: non–negative integer variable representing the number of lot sizes of SKU $i \in \mathscr{F}$ produced or purchased a time t

5. y_{it}: binary variable equal to 1 if $\hat{x}_{it} > 0$ and to 0 otherwise, used to guarantee that $\hat{x}_{it} \geq LS_i$ for SKU $i \in \mathscr{S}$ whenever i is produced or purchased
6. $\hat{y}_{it}$: binary variable equal to 1 if SKU $i = 2, 3, \ldots, P$ is used at time t and to 0 otherwise, used to guarantee that the minimal production quantity F_i is produced whenever i is used.

Then, we formulated the following mixed integer linear programming model:

$$\min \sum_{i=1}^{P} \sum_{t=1}^{T} (T - t + 1) x_{it} \tag{6}$$

$$I_{1t} = \bar{I}_1 + \sum_{\tau=1}^{t-LT(1)} x_{1\tau} - \sum_{\tau=1}^{t} (D_\tau + C_\tau) \qquad t = 1, 2, \ldots, T \tag{7}$$

$$I_{it} = \bar{I}_i + \sum_{\tau=1}^{t-LT(i)} x_{i\tau} - \sum_{\tau=1}^{t} \left(\sum_{j=1}^{P} R_{ij} x_{j\tau} + F_i \hat{y}_{it} + IA_{i\tau} + FA_{i\tau} \right) \quad i = 2, 3, \ldots, P \quad t = 1, 2, \ldots, T \tag{8}$$

$$x_{1t} = \hat{x}_{1t} + S_t \qquad t = 1, 2, \ldots, T \tag{9}$$

$$x_{1t} \leq M\pi_t \qquad t = 1, 2, \ldots, T \tag{10}$$

$$x_{it} = \hat{x}_{it} \qquad i = 2, 3, \ldots, P \quad t = 1, 2, \ldots, T \tag{11}$$

$$\sum_{j=1}^{P} R_{ij} x_{jt} \leq M \hat{y}_{it} \qquad i = 2, 3, \ldots, P \quad t = 1, 2, \ldots, T \tag{12}$$

$$\hat{x}_{it} = LS_i z_{it} \qquad i \in \mathscr{F} \quad t = 1, 2, \ldots, T \tag{13}$$

$$\hat{x}_{it} \geq LS_i y_{it} \qquad i \in \mathscr{S} \quad t = 1, 2, \ldots, T \tag{14}$$

$$\hat{x}_{it} \leq M y_{it} \qquad i = 1, 2, \ldots, P \quad t = 1, 2, \ldots, T \tag{15}$$

$$x_{it} \geq 0 \qquad i = 1, 2, \ldots, P \quad t = 1, 2, \ldots, T \tag{16}$$

$$\hat{x}_{it} \geq 0 \qquad i = 1, 2, \ldots, P \quad t = 1, 2, \ldots, T \tag{17}$$

$$I_{it} \geq 0 \qquad i = 1, 2, \ldots, P \quad t = 1, 2, \ldots, T \tag{18}$$

$$y_{it} \in \{0, 1\} \qquad i = 1, 2, \ldots, P \quad t = 1, 2, \ldots, T \tag{19}$$

$$\hat{y}_{it} \in \{0, 1\} \qquad i = 2, \ldots, P \quad t = 1, 2, \ldots, T \tag{20}$$

$$z_{it} \geq 0 \text{ integer} \quad i = 1, 2, \ldots, P \quad t = 1, 2, \ldots, T \tag{21}$$

The objective function (6) penalizes the production or the purchase of SKUs performed before the time bucket in which these quantities are needed (delivery for SKU 1 and next steps in production for the other SKUs). Constraints (7) compute the inventory level of the final product at the end of each time bucket as the sum of the starting inventory level (I_1), plus the total quantity produced up to time $t - LT_1$ (and therefore available at time t), minus the sum of the cumulative demand and of the planned deliveries up to time t. Constraints (8) compute the inventory level of each other SKU i at the end of each time bucket, as the sum of the starting inventory level I_i, plus the total quantity produced up to time $t - LT_i$, minus the cumulative demand, given by the sum of the number of units needed to produce other SKUs, the minimum quantity to produce whenever the SKU i is used, the number of units of SKU i used at time bucket t to produce other final products and the number of units of SKU i scheduled at time t to produce other final products. Constraints (9) compute the quantity of the final product produced at each time bucket t as the sum of the net quantity and of the quantity to produce at time t from previous planning. Constraints (10) guarantee that the production of the final product can be performed only when it is allowed. Constraints (11) define the quantity of each SKU different than the final product equal to the net quantity. Constraints (12) guarantee that if a SKU is used at time bucket t, then the minimal number of units of the SKU is produced. Constraints (13)–(15) guarantee that, for each time bucket at which a SKU is produced or purchased, the produced or purchased quantity at each time bucket is a multiple of the lot size for the SKUs in the set $\mathscr{F}$ and at least equal to the corresponding lot size for the SKUs in the set $\mathscr{S}$. Constraints (16)–(21) define the decision variables of the problem. Note that Constraints (18) also guarantee that no stock-out can occur for each SKU.

This model is a new variant of the one introduced in [3]. It was implemented by using the modeling language AMPL in the Solver Studio framework embedded in Excel [7]. We also formulated a second variant of the model able to guarantee that the inventory level of each SKU at the end of each time bucket is enough to satisfy the demand of the SKU for a given number of time buckets.

5 Computational Results

In this section we provide, as an example, some of the computational results obtained, by applying the suggested forecasting methods and optimization models to the instances provided by the Company.

5.1 Forecasting Methods

In this section we describe how to predict sales data for the 200 different products produced by the Company (including third-party brands). We have at our disposal weekly/bi-weekly selling data starting from January 2013.

The analysis of each product was splitted into two parts: We first analyzed the historical time series by the models described in Sect. 3 (LOESS and HW) with the aim to identify the best method taking into account the seasonality and the presence of trend. The second part focused on the prediction of each product by choosing the best method between LOESS and HW based on the lowest mean squared error of the forecasts obtained by removing the last 26 observations with the 26 observations themselves.

The analysis has been performed in R [6], a free software environment for statistical computing and graphics. The tool takes as input the historical series of all the products and creates as output a worksheet containing for each product the predictive method used (i.e. LOESS or HW) and the demand forecast for the next six future observations. The tool can be used autonomously by the Company, updating the input file week after week with the new observations and recalculating the future forecasts. Notice that the instrument will become more and more reliable over time handling longer historical series.

Let us now look in detail at the analysis carried out for each of the products. For the sake of brevity, we will limit the analysis to a particular type of yogurt identified as 3073 FPLUS BIANCO 600G CLV. The same methodology has been applied for all the remaining products.

Figure 2 show the decomposition of the series of yogurt 3073 FPLUS BIANCO 600G CLV with the LOESS model.

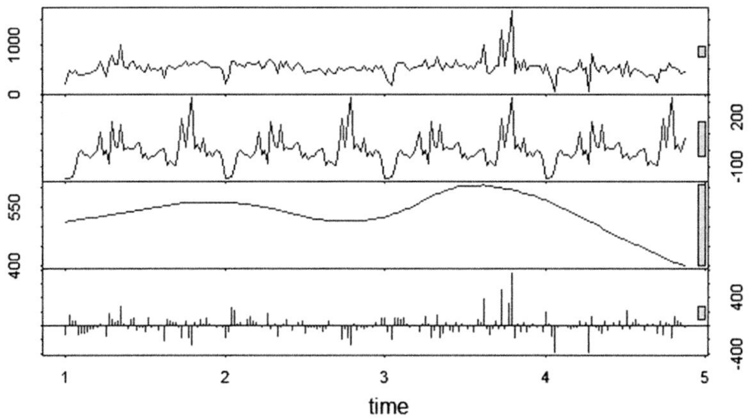

Fig. 2 LOESS decomposition for the product FPLUS BIANCO 600G CLV

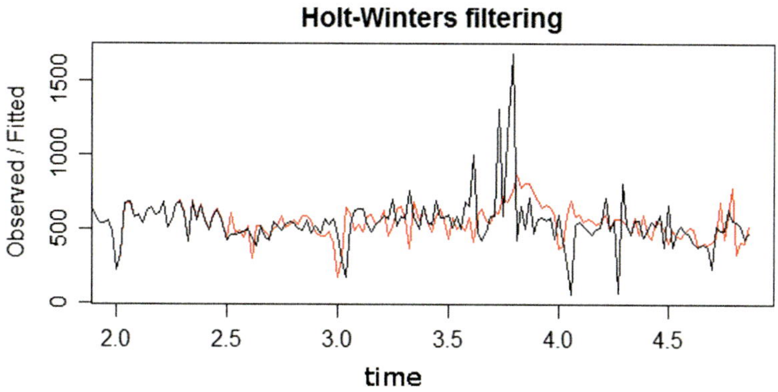

Fig. 3 HW modelling for the product FPLUS BIANCO 600G CLV

The first row shows the historical series of the last five years, the second the seasonality, the third the trend decreasing in the last period and in the fourth row the residuals.

On the other hand, Fig. 3 shows a comparison between the original historical series (in black) with the Holt-Winters model (in red) over the last five years.

In order to identify which one of the two considered models best describes the trend of the series, we have computed the mean square error (MSE) and analyzed the Q-Qnormal Plot, which is constructed with the quantiles of the empirical distribution of the residues and the quantiles of the standard normal distribution.

From the analysis of the Q-Qnormal Plot the best method seems to be LOESS, even if the choice between the two methods turns out to be difficult because, in both cases, the graphical representation of the cumulative distribution of the observed variable seems to follow the cumulative distribution of the normal distribution. In order to determine the best method to describe the series, we therefore calculated the respective mean square errors (MSE): $MSE\ LOESS = 14848.43$, $MSE\ HW = 30504.33$. Since the mean square error of the LOESS method is the smallest one, we can therefore conclude that LOESS is the method that best describes the trend of the series of the product analyzed.

We finally decided which of the LOESS and HW methods is the best for a prediction. Given the historical series in which we remove the last 26 observations, we generated for these 26 observations a prediction with LOESS and HW and finally we chose the best method for a forecast between the two, by calculating their mean square errors (MSE).

Figure 4 refers to the LOESS prediction: In black the original series is represented, while in blue the LOESS forecast up to 2018. The overlapped part is the part we compared (26 observations) and the LOESS mean square error is $MSE = 13898.83$, while the one of HW prediction is $MSE = 59863.83$.

We can therefore conclude that LOESS turns out to be the best forecasting methods for FPLUS BIANCO 600G CLT.

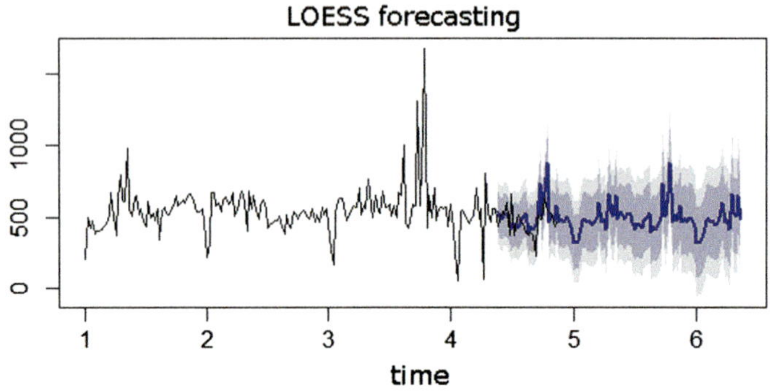

Fig. 4 LOESS forecasting: in black the original series while in blue the LOESS forecast up to 2018

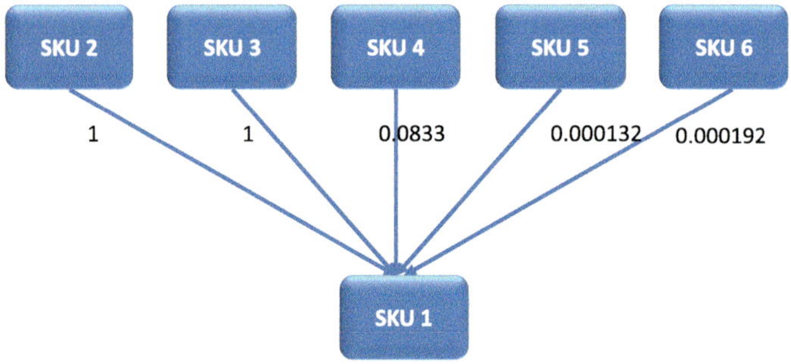

Fig. 5 Bill of Materials (BOM)

The application of this methodology to all the 200 products has allowed to the Company to reduce the average percentage error on a 6 weeks production schedule from 29.71% to only 7.94%.

5.2 An Optimized Material Requirements Planning (MRP)

We now provide an example of the application of the MRP model formulated in Sect. 4 to a final product (a UHT milk) produced by the Company, having the following data: Number of SKU P: 6 (final product: SKU 1, components: SKU 2, SKU 3, SKU 4, SKU 5, SKU 6); number of time buckets T: 42 days. Figure 5 shows the Bill of Materials (BOM) of this final product: One unit of the final product SKU 1 is obtained by assembling SKU 2, SKU 3, SKU 4, SKU 5, SKU 6 in the quantities R_{ij} reported on the BOM.

The demand of the final product over 6 weeks (from July, 11 to August, 21 in 2016) is the following (the demand is 0 in the days not included in the table):

Day t	1	8	15	22	29	36
Demand D_t	15276	68364	84900	17820	17820	17820

The MRP requires, for each SKU, the following data:

SKU i	Lead time LT_i	Lot size LS_i	Lot size set	Starting inventory level $\bar{I}_i$
1	0	1000	$\mathscr{F}$ (multiple)	153475.00
2	28	1440000	$\mathscr{S}$ (minimum)	1024928.00
3	28	16200	$\mathscr{F}$ (multiple)	611547.00
4	28	10000	$\mathscr{S}$ (minimum)	9236.00
5	21	675	$\mathscr{S}$ (minimum)	636.33
6	28	384	$\mathscr{S}$ (minimum)	561.61

The application of the model formulated in Sect. 4 to this case showed that, if the production and purchase quantities applied by the Company are used in the optimization model, the optimal requirements planning is equal to the one provided by the MRP embedded in the information system of the Company. This states that our application is equivalent to the MRP embedded in the information system, when the same solution is adopted. Moreover, when an optimal solution is computed, a significant reduction in the inventory levels are obtained, as shown in Fig. 6 for the final product, where we compare the inventory levels in the solution used by the Company and in the optimal solution.

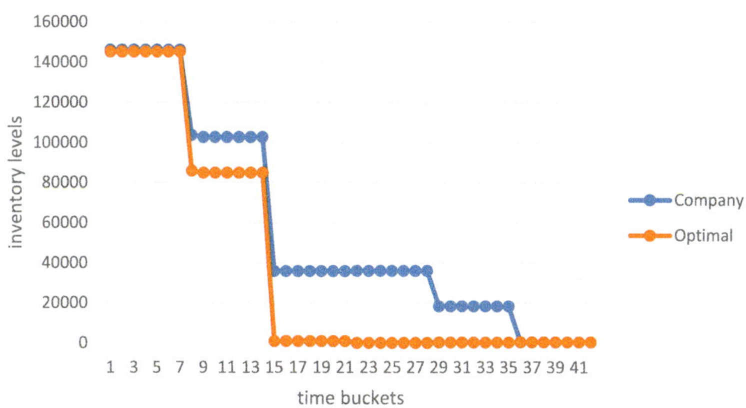

Fig. 6 Inventory levels of the final product

We applied this methodology to the instances provided by the Company, related to the production of two types of final products in 2016. The results showed that the inventory levels can be improved by adopting an optimal solution, similarly to what happened in the example shown before.

6 Conclusions

Operations Research was recently defined as "The Science of Better". This project provides an example of successful application of statistical and Operations Research techniques to a complex real case. However, this project is just a first step. Further efforts could be devoted, for example, in embedding demand forecasts in the MRP optimization models, maybe formulating *Stochastic Programming* models able to compute optimal requirements planning given a set of scenarios of demands.

Acknowledgements The authors would like to thank "La Centrale del Latte di Vicenza SpA", in particular Dr. Andrea Barban and Dr. Gian Luigi Gallio for the description of the problem, the provided historical data and for stimulating this research. The authors would like to thank the "Sportello Matematico per l'Industria Italiana", in particular Dr. Valentina Gratta, Dr. Anna Melchiori and Dr. Antonino Sgalambro, for the support provided in the management of this project. Finally, the authors would like to thank Dr. Elena Poiatti for the support given in the analysis of the data.

References

1. Cleveland, W.S.: Robust locally weighted regression and smoothing scatterplots. J. Am. Stat. Assoc. **74**, 829–836 (1979)
2. Holt, C.C.: Forecasting seasonals and trends by exponentially weighted moving averages. ONR Research Memorandum, Carnegie Institute of Technology 52 (1957). https://doi.org/10.1016/j.ijforecast.2003.09.015
3. Voß, S., Woodruff, D.L.: Introduction to Computational Optimization Models for Production Planning in a Supply Chain. Springer, Berlin (2006)
4. Wack, J.G., Lummus, R.R.: Sales forecasting for strategic resource planning. Int. J. Oper. Prod. Manag. **22**(9), 1014–1031 (2002)
5. Winters, P.R.: Forecasting sales by exponentially weighted moving averages. Manag. Sci. **6**, 324–342 (1960)
6. https://www.r-project.org/
7. https://solverstudio.org/

Production Scheduling and Distribution Planning at FATER S.p.A.

Giovanni Giallombardo, Giovanna Miglionico, Lucia Nitoglia and Sergio Pelle

Abstract We present the relevant outcomes of the research co-operation between Università della Calabria and FATER S.p.A., an Italian nationwide company holding several leading brands in the area of disposable diapers. The aim of such co-operation has been to tackle some issues regarding production scheduling and distribution management at FATER. In particular, two problems have been addressed: (i) the optimal lot-size management of the FIX2 production line located in the Pescara plant, and the distribution of goods from the production plant to the distribution depots covering the entire national territory. As for the former, the classic production-rotation model has been adapted to the FATER framework for taking into account different possible scenarios in terms of manpower weekly shifts. As for the latter, an inventory–routing model adopting a simplified column-generation approach has been proposed. We describe the two optimization problems and the adopted solution approaches, giving also some details on the related software packages implementing such methods, along with the relevant results.

Keywords Lotsize · Setup time · Vehicle routing · Inventory control

G. Giallombardo (✉) · G. Miglionico
Dipartimento di Ingegneria Informatica, Modellistica, Elettronica e Sistemistica (DIMES), Università della Calabria, Rende, Italy
e-mail: giallo@dimes.unical.it

G. Miglionico
e-mail: gmiglionico@dimes.unical.it

L. Nitoglia · S. Pelle
FATER S.p.A., Pescara, Italy
e-mail: nitoglia.l@fatergroup.com

S. Pelle
e-mail: pelle.s@fatergroup.com

M. Dell'Amico et al. (eds.), *A View of Operations Research Applications in Italy, 2018*, AIRO Springer Series 2, https://doi.org/10.1007/978-3-030-25842-9_8

1 Introduction

FATER S.p.A. is a large Italian company, an equal joint venture between Gruppo Angelini and Procter & Gamble since 1992, and it is leader in the market of absorbent personal products. Established in Pescara (Italy) in 1958 by Gruppo Angelini, FATER has been in the italian houses for 60 years through several well known products and brands: baby napkins and diapers (Pampers brand), feminine pads and panty-liners (Lines brand), internal tampons (Tampax brand), and products for incontinence (Lines Specialist brand). In 2013 FATER acquired the ACE bleach and home detergent brand that is distributed in 39 countries scattered over Western Europe and Ceemea markets. FATER headquarter is located in Pescara together with the production plant of absorbent products; other plants are in Campochiaro (Italy), Gebze (Turkey), Mohammedia (Morocco), Porto (Portugal). Legal addresses of FATER are also in Portugal, Morocco, Turkey, Romania, and Russia. The company continuous growth is strongly based on a deep knowledge of the consumer behavior combined with working methodologies typical of multinational companies.

Motivated by the company interest in improving the integration between production and distribution, FATER has funded and awarded a research contract to the Dipartimento di Ingegneria Informatica, Modellistica, Elettronica e Sistemistica (DIMES) at Università della Calabria (Italy), with the aim of reducing the costs associated to stock levels at both the production plant and the distribution depots.

In this chapter we survey the relevant outcomes of the co-operation between DIMES and FATER. In particular, in Sect. 2 we introduce the problem faced by FATER in managing its FIX2 production line, together with the optimization models used to calculate the lot size with and without synchronization of all productions. The models are also extended to the case of non-zero setup times, and the software package `ProdOpt` is introduced, which implements the optimization models for the lot size management. In Sect. 3, we present the distribution problems faced by FATER in transferring goods from the production plant to the distribution depots. Next, the optimization model used to find the best route-cargo configuration, along with the periodicity of each visit to the distribution depots, is presented. Moreover, the software package `RoutingOpt`, implementing the optimization model for the distribution management, is described. Finally, some concluding remarks are given in Sect. 4.

2 Lot Sizing and Production Rotation Management

We address the problem of managing the production line FIX2 at FATER, where six product typologies are manufactured via six different processes. Since the production line is a shared resource, each process takes place in lots by adopting a rotation policy which assigns appropriate time slots to the production of the different items. In such a setting, particular attention has to be paid to the setup (or changeover) costs which are incurred every time the production of a certain item is stopped, and the line has to be adapted to accommodate the production of yet another type of item. Hence, the cost dynamics can be easily described by noting that small lots correspond to small inventory cost and large setup cost per time-unit, whereas the reverse is true in case of large lots (i.e., large inventory cost and small setup cost per time-unit).

The problem clearly belongs to the wide class of inventory/production management and control problems for which a huge body of related literature exists. For the scope of this article, here we only mention the seminal works available in [2] for inventory/production management motivations, and the description of several optimization models and the related solution approaches, while an interesting survey on models and algorithms for capacitated lot sizing problems can be found in [3].

We start by presenting the easiest version of the problem, where no setup time is accounted for in the formulation. From the mathematical point of view, the problem data are listed in the following, where for notation simplicity we indicate by "product" any of the six "product typologies" previously mentioned. We denote by $J \triangleq \{1, \ldots, n\}$ the index set of products, where $n = 6$ in the FATER production process. Each product $j \in J$ is produced at a rate of P_j units per time-unit, and demanded at a rate D_j units per time-unit. Moreover, we denote by F_j and h_j, respectively, the fixed setup cost and the unit inventory carrying cost per time-unit of product $j \in J$. We remark that a necessary condition for problem feasibility is that the overall utilization coefficient of the line, evaluated independently of the setup times, is smaller than one, that is

$$\sum_{j \in J} \frac{D_j}{P_j} < 1, \tag{1}$$

where the overall utilization coefficient is the sum of the individual utilization coefficients of the various items, as D_j/P_j represents the fraction of total time the line is to be devoted to production of item $j \in J$. Strict inequality in formula (1) is justified by the observation that equality would mean full utilization of the production line, which results in no possibility of accommodating any setup time (see the discussion in Sect. 2.1). We will see in the following that calculation of such coefficients deserves some attention in our application.

In order to formulate the problem we have adopted an extension of the classical single-product economic order quantity (EOQ) model [5, Chap. 3] to the single-facility multi-product system with finite replenishment rates, see [2]. The objective is to minimize the total production and inventory costs per time-unit acting on the decision variables Q_j, the lot-size of item $j \in J$ to be produced periodically, and T_j, the cycle length of item $j \in J$, namely, the time interval between two successive starts of the production of item j. We remark that Q_j and T_j are not independent of each other, as the following relation, coming from mass-balance, holds:

$$Q_j = D_j T_j, \quad \forall j \in J. \tag{2}$$

As a consequence, any of the two groups of variables, lot size or cycle length, can be equivalently taken as the unique group of decision variables.

Next, we define the objective function $K(Q_1, \ldots, Q_n)$, depending on the lot size of every item, as the total production- and inventory-cost per time-unit [2, Sects. 3–6]

$$K(Q_1, \ldots, Q_n) \triangleq \sum_{i \in J} \left(\frac{F_j D_j}{Q_j} + h_j \frac{Q_j}{2} \left(1 - \frac{D_j}{P_j} \right) \right). \tag{3}$$

In particular, the total cost per time-unit $K(Q_1, \ldots, Q_n)$ is the sum of n independent single product costs. In fact, the ratio $\frac{D_j}{P_j}$ is the number of lots per time units and

$$\frac{Q_j}{2} \left(1 - \frac{D_j}{P_j} \right) \tag{4}$$

is the average inventory of item $j \in J$ per time-unit, as a function of the lot size Q_j under the hypothesis of stationary demand rate. We note that variable production costs are not accounted for in (3) since their per unit-time contribution is constant, namely, not dependent on the lot sizes. We also remark that in this classic formula the conflicting effects of any variation in the size of a lot are well represented. In fact, any increase in such size produces a beneficial effects on the setup costs, while increasing the inventory costs. Of course the opposite happens in case of lot size reduction.

Although $K(\cdot)$ appears as a function of as many as n variables, it is immediate to observe that it can be represented as the sum of n functions of the single variables $Q_1, \ldots, Q_n$. Thus, the optimal solution vector can be obtained by putting together the optimal scalar solution $\widetilde{Q}_j$ of each such functions. The following classic EOQ formula holds, which can be obtained by letting equal to zero the derivative of each single-variable function:

$$\widetilde{Q}_j = \sqrt{\frac{2F_j D_j}{h_j \left(1 - \frac{D_j}{P_j} \right)}}. \tag{5}$$

For later convenience, we observe that, as soon as setup times will be accounted for, each lot-size cannot be reduced below a given threshold. In such a case, a lower-bound constraint on the lot-size of the type

$$Q_j \geq Q_j^{min}, \quad \forall j \in J, \tag{6}$$

can be easily taken into account, thanks to convexity of the functions involved, by letting

$$\widetilde{Q}_j = \max \left\{ Q_j^{min}, \sqrt{\frac{2F_j D_j}{h_j \left(1 - \frac{D_j}{P_j} \right)}} \right\}, \quad \forall j \in J. \tag{7}$$

The above calculations are interesting because they provide an insight on which would be the optimal lot sizes for all items, should they be considered in a "stand alone" setting. Unfortunately, such solution is not always feasible in practice as the model does not account for the shared resource represented by the production line. Consequently, the productions would lack of synchronization, giving rise to possible delays in some productions, with consequent shortages.

A possible way to overcome such difficulty is to choose the same cycle length for the production of all items. Such choice, indeed, allows complete synchronization of all productions, thus avoiding possible conflicts in the use of the line. By imposing the condition

$$Q_j = T D_j, \quad \forall j \in J \tag{8}$$

in (3), the new model can be easily obtained as

$$K(T) = \frac{1}{T} \sum_{j \in J} F_j + \frac{T}{2} \sum_{j \in J} h_j D_j \left(1 - \frac{D_j}{P_j}\right), \tag{9}$$

where the updated cost function depends only on the common cycle length T.

Also in this case the (convex) cost function can be easily minimized in closed form by letting the derivative equal to zero, thus obtaining T^*, the optimal cycle length, as

$$T^* = \sqrt{\frac{2 \sum_{j=1}^{n} F_j}{\sum_{j=1}^{n} h_j D_j \left(1 - \frac{D_j}{P_j}\right)}}, \tag{10}$$

which allows to calculate the optimal synchronized lot-size Q_j^* as

$$Q_j^* = T^* D_j, \quad \forall j \in J. \tag{11}$$

Formulas (10) and (11) solve the problem known in the literature [2] as the single-facility lot-size model.

2.1 Setup Time Management

As previously remarked, in order to enhance the formulation of the problem we need to account for setup times, the obvious approach being to prevent the selection of arbitrarily small cycle length and, consequently, too fragmented lot sizes.

In the following we show how the minimal threshold on the cycle length can be determined, by first assuming that setup times are not "sequence dependent", namely, that the setup time of item $j \in J$, say s_j, does not depend on the item produced immediately before it. Denoting by τ_j the time needed to produce a lot of size Q_j of item $j \in J$, solution feasibility requires that the cycle length T is sufficiently large to accommodate both the production and the setup times for all items. Thus, the following inequality must hold:

$$\sum_{j \in J} \tau_j + \sum_{j \in J} s_j \leq T. \tag{12}$$

From

$$\tau_j = \frac{Q_j}{P_j}, \tag{13}$$

which expresses the production time as a function of the lot size and the production rate, and taking into account (8) we obtain the following inequality

$$T \geq \frac{\sum_{j \in J} s_j}{1 - \sum_{j \in J} \frac{D_j}{P_j}} \triangleq T_{min} \tag{14}$$

which provides the lower threshold T_{min} for any cycle length T to be physically feasible. We remark that the higher is the global utilization coefficient $\sum_{j \in J} D_j / P_j$, the higher (i.e., more restrictive) is T_{min}. It is also obvious that large values of the setup times enforce large T_{min}, which, in turn, results in large lot sizes and inventory costs. As a consequence, it is necessary to define the optimal constrained cycle length by solving the constrained problem

$$\min\{K(T) : T \geq T_{min}\} \tag{15}$$

whose solution, due to convexity of $K(T)$, offers no difficulty, returning the optimal constrained cycle-length $\overline{T}^*$ as

$$\overline{T}^* = \max\{T_{min}, T^*\} \tag{16}$$

and the optimal constrained synchronized lot-size $\overline{Q}_j^*$ as

$$\overline{Q}_j^* = \overline{T}^* D_j \quad \forall j \in J. \tag{17}$$

It is worth noting that calculation of T_{min} is well posed only if condition (1) holds true. In the FATER application considered here, such condition is fulfilled since the production plan is usually spanned on a mix of two or three eight-hour shifts per working day. In fact, from the demand and production rates described in the next section, the global utilization coefficients can be calculated as 1.84, 0.92 and 0.61 for one, two and three eight-hour shifts, respectively. Hence, any weekly production

Table 1 Model data and their actual unit of measure

Production rate	Number of pallets per hour
Demand rate	Number of pallets per working week (5 days)
Inventory cost	Euro per pallet per month
Setup time	Hour per lot couple
Setup cost	Euro per lot couple

plan based on a mix of two and three eight-hour shifts brings to a global utilization coefficient in the feasible range 61–92%, while condition (1) would not be satisfied should the production take place during only one eight-hour shift per working day.

Finally, we further enhance our formulation by accounting for sequence dependent setup times. Denoting by s_{ij}, with $i, j \in J$, the time necessary to prepare the production of item j following the production of item i, the way in which the items are sequenced (and thus the setup times spent) inside the production cycle becomes relevant. Looking for the best sequence amounts to construct a complete directed graph characterized by a number of nodes equal to n, the total number of items. The weight of arc (i, j), $i, j \in J$, $i \neq j$, is s_{ij} and the sequence which minimizes the total setup time is the one obtained by solving an asymmetric Traveling Salesman Problem (TSP) on the graph. Note that in our application the number of product typologies is small and solving the TSP is not at all difficult from the computational point of view. Once L_{min}, the length of the minimal Hamiltonian circuit on the graph, has been found, the threshold T_{min} on the cycle length becomes

$$T_{min} = \frac{L_{min}}{1 - \sum_{j \in J} \frac{D_j}{P_j}}. \tag{18}$$

2.2 Practical Implementation: The `ProdOpt` Package

Every optimization model presented above has been implemented in C# and embedded in a software package named `ProdOpt`. In order to have the models work properly, we have made some adjustments, in terms of units of measure, of the rough data collected at FATER. Recalling that there are 6 product typologies, we report in Table 1 the unit of measure of all data needed to construct the models, according to the standards adopted at FATER.

In designing and implementing `ProdOpt`, the production rates have been maintained as number of pallets per hour, and the demand rates, usually expressed in number of pallets per week, have been converted into the same unit of measure. The conversion factor adopted in this case is such that the weekly demand is spanned over a five-day working period. The inventory cost per item unit and per time unit, usually expressed as Euro per pallet per month, has been converted into Euro per pallet per hour. The standard working month has been considered as 28 days long.

As a consequence of the sequence-dependent setup times, the available setup costs are actually sequence dependent too. This forced us to set in the formulas discussed previously

$$F_j = F_{i_j,j}, \quad \forall j \in J, \tag{19}$$

where i_j is the index of the item preceding j in the optimal item sequence obtained by solving the TSP and $F_{i_j,j}$ is the corresponding sequence-dependent setup cost.

The choice of the hour as the time unit has been motivated by the difficulty of evaluating week productivity, which in fact depends on the structure adopted for the time shift in each specific working day. However, `ProdOpt` returns as an output the cycle length expressed in weeks, assuming implicitly that the line works under a three eight-hour working shift, five days per week. Since the time cycle is also returned as an output in terms of number of hours, its resizing in terms of weeks can be easily obtained in case a different shift structure is adopted.

In our implementation, despite of the small size of the actual problem, we have adopted a standard multi-start nearest-neighbor heuristics as a TSP solution algorithm, this allowing to possibly extend the range of applicability of `ProdOpt` even to frameworks characterized by a large number of product typologies. However, as an aside feature, `ProdOpt` allows the user to define the item sequence off the line as well. Of course, in this case, the setup costs are calculated according to such sequence.

3 Distribution Management

As mentioned in the Introduction, production processes at FATER take place at the Pescara plant, where the central warehouse is located too. From there, items are delivered to 22 distribution depots, scattered over the Italian territory, which in turn serve end customers. Products are packaged in pallets, and a fleet of vehicles is employed to reach the depots. Two types of trucks are available, carrying 38 and 47 pallets, respectively.

The research has been motivated by the observation that the "point-to-point" distribution policy adopted at FATER implies that each truck visits exactly one depot per trip. In this case, since the distribution is obviously managed according to a *truckload shipping* policy, an apparent shortcoming is caused by possibly oversizing the stocks at some depots, especially the ones that are more rarely served. Therefore, we have devised a new distribution model, based on a little more complex truck routing, that accommodates to serve two or more depots by the same truck trip. It is worth noting that the number of depots is relatively small, and their geographical distribution suggests that more complex routes are of no interest.

We particularly emphasize that our objective has not merely been to reduce transportation costs. In fact, we have pursued the design of a policy which, in the spirit of the inventory-routing optimization, would allow (i) to control the transportation costs, mainly acting on the empty-truck part of the routes by aggregating two or more depots, and (ii) to reduce the inventory costs at depots by delivering smaller amounts more frequently.

We have approached the problem by first partitioning the depots in a clever way, so that it would appear viable to serve each of the depot clusters by exactly one truck route. In fact, such procedure resembles the well known heuristics for the Vehicle Routing Problem (VRP) based on the so-called *seed points*, which works on a "cluster first, route second" basis (a review of the huge literature on truck routing is outside the scope of the paper, here we only mention the excellent survey [4] and, for periodic routing, [1]). Once the clusters have been fixed, the problem reduces to the (simple) definition of the route, along with the partition of the payload among the depots belonging to the route, and to the calculation of the cycle length, that is the time interval between two successive executions of the route.

We assume that a planning horizon, say one month, is given and that the demand rate D_i (measured in number of pallets per month) is known for each depot i. The inventory holding cost per pallet per month, denoted by H, is assumed known too and identical for all depots. The rationale underlying our approach is based on a simplified logic of the "column generation" type. Before giving a formal description of the approach we introduce an example.

Consider a cluster V formed by three depots denoted as V_1, V_2 and V_3, respectively. All the possible alternative services are described in Table 2, where the depot-route incidence matrix is given, namely, all the seven possible combinations, $R_1, \ldots, R_7$, involving one, two, or three depots are listed (each combination is then implicitly associated to an optimal route in terms of the total traveled distance).

Take now, for an example, column 5 in Table 2, which represents the joint service of depots V_1 and V_3. The payload configurations associated to such column are all the triplets $(a_{15}, 0, a_{35})$ satisfying either

$$a_{15} + a_{35} = 38$$

or

$$a_{15} + a_{35} = 47,$$

according to the type of truck considered, where a_{ij} denotes the number of pallets delivered to depot $V_i \in V$ by a truck traveling the route R_j.

Hence, in order to introduce a more formal setting, given any subset V of the distribution depots, with $|V| = m$, we define the matrix A of the feasible payload

Table 2 Depot-route matrix

	R_1	R_2	R_3	R_4	R_5	R_6	R_7
V_1	1	0	0	1	1	0	1
V_2	0	1	0	1	0	1	1
V_3	0	0	1	0	1	1	1

configurations. Columns of A are those m-dimensional nonnegative integer vectors $\mathbf{A}_j \in \mathbb{Z}_+^m$ (not necessarily to be enumerated!) such that

$$\sum_{i=1}^{m} a_{ij} = 38 \quad \text{or} \quad \sum_{i=1}^{m} a_{ij} = 47. \tag{20}$$

Each column $\mathbf{A}_j$ of A is associated to a subset of depots actually served, say $V^{(j)} \subseteq V$, namely, those depots for which $a_{ij} > 0$. The cost w_j of column $\mathbf{A}_j$ is expressed as:

$$w_j = r_j + h_j, \tag{21}$$

where r_j is the optimal routing cost of the VRP defined on the set $\{V_0\} \cup V^{(j)}$, with V_0 denoting the central warehouse, and h_j is a measure of the inventory costs induced by the payload configuration $\mathbf{A}_j$.

To evaluate h_j we proceed as follows. Since a_{ij} is the payload devoted to depot V_i in the configuration $\mathbf{A}_j$, it holds that

$$t_{ij} = \frac{a_{ij}}{D_i} \tag{22}$$

represents the cycle time, that is the time interval between to successive replenishments of depot V_i when the delivery size is a_{ij}. In such an interval the average inventory at depot $V_i \in V^{(j)}$ is $\frac{a_{ij}}{2}$. Thus, we can define the inventory cost h_{ij} associated to the payload a_{ij} as

$$h_{ij} = \frac{H t_{ij} a_{ij}}{2} = \frac{H(a_{ij})^2}{2D_i}, \tag{23}$$

and finally we get

$$h_j = \sum_{i:a_{ij}>0} h_{ij}. \tag{24}$$

Grouping in the vector d the demands of the subset of depots V, letting J be the index set of all the possible configurations associated to V, and defining the decision variables x_j as the number of times configuration j is used for the depot cluster V, we obtain the problem:

$$\begin{aligned} z_P = \min & \sum_{j \in J} w_j x_j \\ & \sum_{j \in J} \mathbf{A}_j x_j \geq d \\ & x_j \geq 0, \quad \forall j \in J. \end{aligned} \tag{P}$$

Summing up, the optimization problem (P) is to be formulated and solved for each depot cluster V. As previously mentioned, the approach is inspired by the column generation method for Linear Programming, with the significant advantage, in this application, that the number of configurations is sufficiently small to allow early enumeration of all payload configurations. Of course, if necessary, it would even be possible to work out heuristically just a subset of such configurations. We remark that problem (P) should be treated as an integer linear program due to the physical meaning of the variables x_j. Nevertheless, we have treated the problem as a continuous linear program. In fact, in the practical application of the model, feasible solutions are characterized by large values of their components (the number of visits per depot per year typically reaches a few hundred), which makes acceptable rounding up the possible fractional values of the variables, as confirmed by the computational experience reported next (see Table 5).

3.1 Practical Implementation: The `RoutingOpt` *Package*

The approach presented above has been implemented in C# and embedded into a software package named `RoutingOpt`, where each linear program (P) is solved by means of the Microsoft Solver Foundation Library. The relevant features of the original distribution network at FATER are reported in Table 3. We assume that the variation cost due to different vehicle capacity is negligible. Further input data necessary to run `RoutingOpt` are the depot clusters, the optimal routing costs, the cluster demands, and the inventory costs. After analyzing the network we have identified the 11 depot clusters which are reported in Table 4. The clusters have been heuristically constructed on the basis of proximity, accessibility, and demand characteristics. Calculation of the optimal routing costs, for each subset of depots belonging to the cluster, can be easily made since the number of possible routes for each cluster is sufficiently small (no cluster is made up of more that four depots). `RoutingOpt` returns as output data the most convenient route-cargo configurations, and the number of times each route-cargo configuration should be executed over the time horizon. The computational behavior of the continuous linear program (P), compared against its integrality-contrained counterpart, say (IP), has been analyzed by solving both (P) and (IP) for several time spans, ranging from 1 month to 12 months. A summary of the results, for the non-trivial clusters $C06$, $C07$, and $C10$, is presented in Table 5. In particular, we report the best-, the average-, and the worst-values (over the different time spans) of z_P and of the objective function evaluated at its rounded up solution, say z_R, both scaled to 100 with respect to z_{IP} (i.e., a value of 100 means that the obtained solution is optimal for (IP)).

Table 3 Original features of the distribution network

Depot code	Daily average demand (expressed in number of pallets)	Point-to-point route cost (expressed in Euro per route)
V01	17	500
V02	62	500
V03	16	1000
V04	3	1000
V05	6	1500
V06	21	1000
V07	37	1000
V08	5	1000
V09	24	2500
V10	22	1000
V11	22	2500
V12	15	2500
V13	28	1000
V14	14	1000
V15	182	500
V16	9	500
V17	1	500
V18	6	500
V19	12	500
V20	15	1000
V21	5	1000
V22	29	1000

Table 4 Depot clusters features

Cluster code	Assigned depots	Cluster daily demand	Number of candidate routes
C01	V09, V11	46	3
C02	V15	182	1
C03	V05, V12	21	3
C04	V07, V08	42	3
C05	V06	21	1
C06	V01, V02, V16	88	7
C07	V17, V18, V19	19	7
C08	V10	22	1
C09	V13, V14	42	3
C10	V03, V04, V20, V21	39	15
C11	V22	29	1

Table 5 Performance of LP relaxed (z_P) and rounded-up (z_R) solutions

Cluster code	Scaled z_P			Scaled z_R		
	Best	Avg	Worst	Best	Avg	Worst
C06	99.98	99.71	98.55	100.00	100.28	100.88
C07	99.88	98.62	93.30	100.00	101.15	104.00
C10	99.95	99.44	98.65	100.00	101.11	104.54

4 Industrial Follow-Up and Concluding Remarks

We have considered two problems arising in the management of production and distribution at FATER. The production scheduling problem is related to the lot-size management of the FIX2 production line located in the Pescara plant. We have tailored to the specific case a single-facility multi-product model, often referred to as a production-rotation one, that can account for different possible scenarios in terms of manpower weekly shifts. The related software package `ProdOpt` has been installed at Production Planning division of FATER and has been used for calculating the lotsizes for line FIX2 under different configurations of data (demand, manpower shifts, etc.). The main result has been the refinement of the length of the rotation cycles (of about 10%), which has lead to smaller lotsizes with the consequent reduction of the inventories. The distribution management problem is about delivering goods to distribution depots from the central production plant. An inventory–routing model adopting a simplified column-generation approach has been proposed. The related software package `RoutingOpt` has been installed at FATER with the aim of supporting decision-makers in evaluating the option of changing the adopted point-to-point distribution policy. The results have demonstrated that such change is not economically justified, since the reduction obtainable in terms of inventory costs does not balance the increase of transportation costs.

References

1. Gaudioso, M., Paletta, G.: A heuristic for the periodic vehicle routing problem. Transp. Sci. **26**(2), 86–92 (1992)
2. Johnson, L.A., Montgomery, D.C.: Operations Research in Production Planning, Scheduling, and Inventory Control. Wiley, New York (1974)
3. Karimi, B., Fatemi, G., Seyyed, M.T., Wilson, J.M.: The capacitated lot sizing problem: a review of models and algorithms. Omega **31**(2), 365–378 (2003)
4. Laporte, G.: Fifty years of vehicle routing. Transp. Sci. **43**(4), 408–416 (2009)
5. Waters, D.: Inventory Control and Management. Wiley, Chichester (2003)

Social Applications

IoT Flows: A Network Flow Model Application to Building Evacuation

Claudio Arbib, Mahyar T. Moghaddam and Henry Muccini

Abstract This chapter presents a real-time emergency evacuation handling system based on internet of things (IoT) technologies. The IoT infrastructure has a core computational component that is in charge of minimizing the time necessary to evacuate people from a building. The space and time dimension are discretized according to metrics and models in literature, as well as original methods. The component formulates and solves a linearized, time-indexed flow problem on a network that represents feasible movements of people at a suitable frequency. Accurate parameter setting makes the computational time to solve the model compliant with real-time use. An application of the proposed IoT system and its core algorithm to handle safe evacuation test in Palazzo Camponeschi—a building in L'Aquila (Italy) now and then used for exhibitions—is described, and diverse uses of the methodology are presented.

Keywords Network optimization · Linear programming · Emergency evacuation · IoT

1 Introduction

Despite an emergency can barely be imagined static, actual evacuation plans are still based on static maps with pre-selected routes through which pedestrians should move in case of disaster. This approach has several drawbacks: (*i*) it leads all pedestrians into the same routes, thus concentrating people and creating crowd; (*ii*) it ignores abrupt congestion, obstacles or dangerous areas; (*iii*) it disregards individual behaviors, and does not distinguish special categories (e.g. elderly, children, disabled); (*iv*) it

C. Arbib · M. T. Moghaddam · H. Muccini (✉)
DISIM, University of L'Aquila, Abruzzo, Italy
e-mail: henry.muccini@univaq.it

C. Arbib
e-mail: claudio.arbib@univaq.it

M. T. Moghaddam
e-mail: mahtou@univaq.it

M. Dell'Amico et al. (eds.), *A View of Operations Research Applications in Italy, 2018*, AIRO Springer Series 2, https://doi.org/10.1007/978-3-030-25842-9_9

does not elaborate different training indications for security operators in different scenarios; (v) it does not provide evacuation managers and operators either with a comprehensive understanding of possible scenarios, or with real-time situation-awareness.

In this context, IoT systems can be exploited to support rules for quicker and safer evacuation: by tracking people in a building, possible congestion can be detected, best safety paths can be periodically calculated, and consequently evacuation time under changing emergency conditions can be minimized. On the basis of real-time data obtained from an IoT system, our study indicates that simulation can be advocated as a practical way to assess both minimum evacuation time of buildings and feasibility of evacuation plans. This result is obtained by an analytical model based on a flow optimization problem that can be efficiently solved. The model has a combinatorial nature, as it decomposes both space (building plan) and time dimension into finite elements: *unit cells* and *time slots*. The building topology is then represented as a graph with nodes corresponding to cells, and arcs to connections between adjacent cells. Collected data are used to create a second acyclic digraph, indexed on time, that models all the feasible transitions between adjacent cells at any given time slot, given the current occupancy status of each cell. Minimizing the total evacuation time corresponds then to solve a mathematical program that, in the final refinement, has the form of a linear optimization problem.

In the release here described, the system we propose aims at providing operators with a continuously monitored evaluation of the shortest time required to evacuate people at any time present in a given building. This feature has been approved by the Fire Brigade of the City of l'Aquila as a possible mean to adjust visitor entrance in a public area (in this paper, a gallery indoor space) according to safety conditions. The main system features are: (i) continuous update of solutions, so evacuation guidelines can be adjusted according to visitors positions evolving over time; (ii) continuous path feasibility check, so that unfeasible routes can automatically be discarded; (iii) possibility (not achieved in this paper) of incorporating the analytical model into a mobile app that supports emergency units to evacuate both closed and open spaces.

In this paper, not only we detail the IoT system and the optimization model, but also, testing different scenarios based on real data, we describe the methodology followed to set system parameters and find an efficiency/accuracy compromise that makes system design compliant with real-time specifications. All that given, our overall view, backed up by ongoing work, is to integrate the mathematical model and algorithm here proposed with an *IoT infrastructure* and a *mobile app*.

The paper is organized as follows. Literature is briefly discussed in Sect. 2. Section 3 presents the flow model and Sect. 4 refers on how model parameters should be set up to deal with real cases. The application of the model to a real exhibition venue is presented in Sect. 5 and conclusions are finally drawn in Sect. 6.

2 Literature Review

Evacuation routing problems (ERP) for large scale roads and buildings are complex and subject to various modeling issues. In this domain, a pioneering work is due to Choi et al. [5], who modeled a building evacuation problem by dynamic flow maximization where arc capacities may depend on flows in incident arcs. Although dating back to the Eighties and limited to theoretical analysis, the paper provides in our mind a good starting point and deserves consideration in the light of the progress done in Linear Programming solution tools. Chen et al. [4] propose a flow control algorithm to compute evacuation paths according to building plan and total number of evacuees. The model aims at minimizing total evacuation time while assigning an optimal number of evacuees to each evacuation path. However, as network size increases, the associated problem can no longer be solved in real-time. Recent research bases evacuation planning on a transshipment problem. For instance, Schloter et al. [16] study two classical flow models in order to deal with crowd evacuation planning: one algorithm aims at finding the best transshipment as a convex combination of simple lex-max flows over time; a second one computes earliest arrival transshipments that maximize the flow towards the sink for every point in time. Other papers propose a hybrid optimization-simulation approach. To quote an example, Abdelghany et al. [1] integrates a genetic algorithm with a microscopic pedestrian simulation assignment model. The genetic algorithm looks for an optimal evacuation plan, while simulation guides the search by evaluating the quality of the plans generated.

In fact, one crucial issue addressed by recent literature is the ability of finding good solutions in so short time as required by a practical computational core of a real-time IoT system. Today's solvers can foster this achievement, as they can get easily rid of very large problems in fractions of seconds: on one hand, dealing with more variables helps obtain enough resolution to model the necessary details (in terms of both discretization and non-linearity); on the other hand, quick re-computation allows to cope with data that dynamically change over time.

3 Flow Model for Emergency Handling

The following network construction basically follows [3, 5]. The topology of the building to be evacuated is described by a graph $G = (V, A)$ that in [5] is called *static network*. Nodes of G correspond to the unit cells i obtained by embedding the building into a suitable grid that will be discussed in Sect. 5. In general, cells may have different shapes or sizes: for the purpose of our work what is important is that every cell can approximately be traversed, in any direction, in a single time slot. Cell 0 conventionally represents the outside of the building, or in general a safe place. Safe places can of course be disconnected areas, but as their capacity is assumed large enough to guarantee safety, we will represent them all by a single cell (therefore what we assume about cells traversing time does not apply to cell 0). Arcs of G correspond to passages between adjacent cells: the passage has full capacity if cells share a boundary not interrupted by walls, and a reduced capacity otherwise. With no loss of generality, arcs are supposed directed. Let us denote:

T $\{0, 1, \ldots, \tau\}$, set of unit time slots;
y_i^t state of cell $i \in V$ at time $t \in T$, that is, the number of persons that occupy i at t: this number is a known model parameter for $t = 0$ (in particular, $y_0^0 = 0$) and a decision variable for $t > 0$;
n_i capacity of cell i: it measures the maximum nominal amount of people that i can host at any time (in particular, $n_0 \geq \sum_i y_i^0$); this amount depends on cell shape and size; if cells can be assumed uniform one can set $n_i = n$ for all $i \in V, i \neq 0$.
x_{ij}^t how many persons move from cell i to an adjacent cell j in $(t, t+1]$: this gives the average speed at which the flow proceeds from i to j;
c_{ij} c_{ji} = capacity of the passage between cell i and cell j: this is the maximum amount of people that, independently on how many persons are in cell j, can traverse the passage in the time unit (independence on cell occupancy means neglecting system congestion: we will consider this issue later).

The flow model uses an acyclic digraph D with node set $V \times T$ and arc set

$$E = \{(i, t) \to (j, t+1) : ij \in A, t \in T\}$$

Referred to as τ-*time* or *dynamic network* in [5], D models all the feasible transitions (moves between adjacent cells) that can occur in the building in the time horizon T. Transitions are associated with the x-variables defined above, whereas y-variables define the occupancy of each room (and of the building) from time to time. The x- and y-variables are declared integer and subject to the following constraints:

$$y_j^t - y_j^{t-1} - \sum_{i:ij \in A} x_{ij}^{t-1} + \sum_{i:ji \in A} x_{ji}^{t-1} = 0 \qquad j \in V, t \in T, t > 0 \qquad (1)$$

$$0 \leq x_{ij}^t + x_{ji}^t \leq c_{ij} \qquad t \in T, ij \in A \qquad (2)$$

$$0 \leq y_i^t \leq n_i \qquad t \in T, i \in V \qquad (3)$$

Equation (1) is just a flow conservation law: it expresses the occupancy of cell j at time t as the number y_j^{t-1} of persons present at time $t-1$, augmented of those that during interval $(t-1, t]$ move to j from another cell $i \neq j$, minus those that in the same interval leave cell j for another room $i \neq j$. Box constraints (2), (3) reflect the limited hosting capability of the elements of G.

Maximizing outflow in a given time. To model the relation between time and people outflow, one can try to maximize the number of persons evacuated from the building within τ:

$$\max \quad y_0^\tau \qquad (4)$$

This is the *Max Flow Problem* (MFP) considered in [5]. To find the minimum total evacuation time, one can solve an MFP for different τ, looking for the least value that yields a zero-valued optimal solution. To reduce computation time, this optimal

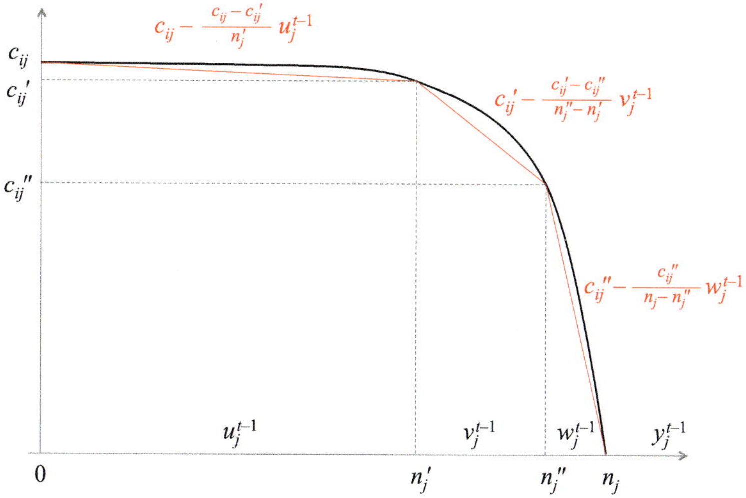

Fig. 1 Congestion curve and a linearization

τ can be computed by logarithmic search. The method can thus provide the decision maker with the Pareto-frontier of the conflicting objectives $\min\{\tau\}$, $\max\{y_0^\tau\}$. The linear structure of the model allows its solution with a large number of variables. Adding variables can help improve model granularity by reducing space and time units (e.g., counting people every 4 s instead of every minute). More importantly, it can also help approximate the non-linearities of arc capacities. In fact, c_{ij} constant in (2) fails to model congestion, that is a situation in which the speed at which the system empties is a decreasing function of room occupancy. A more accurate model of congestion requires arc capacity to be a concave decreasing function of room occupancy. Linearizing this function is quite standard in applications: for simplicity, let us refer to the three-pieces linearization of Fig. 1. One can rewrite

$$y_i^{t-1} = u_i^{t-1} + v_i^{t-1} + w_i^{t-1} \qquad x_{ij}^t = \phi_{ij}^t + \chi_{ij}^t + \psi_{ij}^t \tag{5}$$

with $u_i^{t-1}, v_i^{t-1}, w_i^{t-1}$ non-negative and subject to upper bounds

$$u_i^{t-1} \le n_i' \qquad v_i^{t-1} \le n_i'' - n_i' \qquad w_i^{t-1} \le n_i - n_i' \tag{6}$$

Arc capacity constraints (2) are then replaced by

$$\begin{aligned} 0 &\le \phi_{ij}^t \le c_{ij} - \frac{c_{ij} - c_{ij}'}{n_j'} u_j^{t-1} \qquad (7) \\ 0 &\le \chi_{ij}^t \le c_{ij}' - \frac{c_{ij}' - c_{ij}''}{n_j'' - n_j'} v_j^{t-1} \\ 0 &\le \psi_{ij}^t \le c_{ij}'' - \frac{c_{ij}''}{n_j - n_j''} w_j^{t-1} \end{aligned}$$

Consistency of the ϕ, χ and ψ variables with the x flow variables requires $\chi = 0$ ($\psi = 0$) if ϕ (if χ) does not saturate its capacity. To see this, we rephrase (7):

$$\begin{array}{lllll} 0 & \le & \phi_{ij}^t & \phi_{ij}^t + a_{ij} u_j^{t-1} \le c_{ij} \\ 0 & \le & \chi_{ij}^t & \chi_{ij}^t + a'_{ij} v_j^{t-1} \le c'_{ij} \\ 0 & \le & \psi_{ij}^t & \psi_{ij}^t + a''_{ij} w_j^{t-1} \le c''_{ij} \end{array} \tag{8}$$

where

$$a_{ij} = \frac{c_{ij} - c'_{ij}}{n'_j} \qquad a'_{ij} = \frac{c'_{ij} - c''_{ij}}{n''_j - n'_j} \qquad a''_{ij} = \frac{c''_{ij}}{n_j - n''_j}$$

and $a_{ij} < a'_{ij} < a''_{ij}$. We then observe a simple property hold by optimal basic solutions, that can easily be generalized to any piecewise linear approximation of the congestion curve.

Proposition 3.1 *Suppose that, in a feasible solution, $\bar{u}_i^t < n'_i$ and $\bar{v}_i^t > 0$. Let*

$$\delta = \min\{n'_i - \bar{u}_i^t, \bar{v}_i^t\}$$

Then a solution with $u_i^t = \bar{u}_i^t + \delta$, $v_i^t = \bar{v}_i^t - \delta$ and the other components unchanged is also feasible and no worse than the given one.

Proof See [12].

4 Setting Model Parameters

To get a reliable model, parameters must be set to numbers that reflect reality. Those numbers depend on several considerations, the most relevant being: model granularity, walking velocity in various conditions (on a flat, on staircases etc.), door (and staircase) entrance capacities, room (and staircase) capacities.

4.1 Granularity

The issue of model granularity touches both spatial and temporal units, and affects the shape, size and neighborhood function of the unit cells in which the building is decomposed, as well as the slots that form the evacuation time horizon.

Cell shape. As described in the previous sections we embed the building plan into a grid, whose cells are assumed isometric: that is, can be crossed in any direction in the same amount of time. That amount will define the time slot duration, and cells will be regarded as virtual unit-rooms that communicate one another via physical or virtual (i.e., open space) doors. Grid geometry can vary. Ideal isometric cells are circles, but circles are not embeddable into a grid with adjacent cell sides. Hexagon cells are a good compromise between isometry and plan embedding. However, in our case study we found room sizes and shapes well compatible with a *quasi-square* grid where each room is split into an integer number of cells. Although almost square, those cells are not isometric and therefore transitions in different directions require different time. This implies some caution in constructing the dynamic network D, as it will be explained later.

Cell size. Cell size has an evident effect on the resulting spatial patterns, and consequently on both computational efficiency and model accuracy: the largest the cell, the lesser vertices in G and the lower the refresh frequency at which people positions are updated. Given the speed at which people moves and data are acquired, relatively low refresh frequencies are not an issue; instead, an exact partition of each room in identical cells may result in a huge network. Thus, one can in general approximate the diverse room shapes by $a \times b$ rectangles as large as possible, while minimizing the error introduced. Various ways can be adopted to measure approximation error: the most natural is the difference between real and approximated room area, in which case, for room k of size $p_k \times q_k$, the error is given by

$$e_k(a, b) = q_k[p_k \text{mod}(a)] + p_k[q_k \text{mod}(b)] - [p_k \text{mod}(a)][(q_k \text{mod}(b)]$$

As we need isometric cells and look for a uniform approximation, we set $a = b$ and find a minimizing $\max_k\{e_k(a, a)\}$, meanwhile limiting the total number of approximating cells to some predefined m: hence we choose among the values of a that fulfill $\sum_k \lfloor p_k/a \rfloor \lfloor q_k/a \rfloor \leq m$. A brief description of the method implementation is presented in Sect. 5.

4.2 *Walking Velocity and Door/Cell Capacity*

Walking velocity. The cornerstone on which the length of each unit time slot in T—and consequently its reciprocal, the monitoring frequency—is established, is the *free flow walking velocity*, i.e. the speed at which humans prefer to walk in non-congested and non-hampered conditions. Clearly, its value varies for different categories of people (child, adult, elderly, disable etc.) and slope (flat, upstairs, downstairs). This parameter is important to perceive the distance that an individual can possibly walk during a specific period of time, and its evaluation contributes to define the cells in which an area is to be divided in order to best approximating travel time. Table 1 reports different estimates of pedestrian free flow velocity found in literature.

For our optimization purposes, a single global velocity is needed in homologous areas: in our case, we distinguish flat zones and stairways. For the former, we consider the average free flow walking speed for a flat surface equal to 1.2 m/s (Ye's estimate). For the latter, we observe that in our study the flow is rarely upward and all people

Table 1 Pedestrian free flow velocity

Flat (m/s)		Reference	Stairs (m/s)		Reference
Under 65	Over 65		Up	Down	
1.36		Fruin [8]	0.56	0.65	Fruin [8]
1.36		Weidmann [18]	0.61	0.69	Weidmann [18]
1.25	0.97	Knoblauch et al. [10]	0.8 ± 0.19		Kratchman [11]
1.042–1.508	0.889–1.083	TranSafety Inc. [17]	0.59	0.76	Jiang et al. [9]
1.20		Ye et al. [19]	0.81 ± 0.13		Fang et al. [7]
			0.83	0.86	Patra et al. [15]

located in first floor are supposed to go downstairs to safe places located at the building ground floor: for this situation we stay at Kratchman's estimate that is 0.61 m/s. Therefore, the speed in flat zones is almost twice that of on downward stairways. Different speeds in different building areas will be treated by varying network granularity, as it will be clear in the following; the approach can easily be generalized to diverse situations.

Door capacity. The capacity of a door depends on such various aspects as user composition, door type (always open, open when used, turnstile), crowdedness and, last but not least, door width. A study by Daamen et al. [6] focuses on the relationship between door capacity, user composition and stress level, arguing an average 2.8 p/s for a 1-m width door (p/m/s). From literature review, door capacities turn out in a range between 1.03 and 3.23 P/m/s. None of the aforementioned studies consider bidirectional flow through doors, and so do we in our optimization model: in fact, although bidirectional flows are in principle possible, they will never occur in an optimal solution even without considering flow rate reductions due to collisions. Taking advantage of the aforementioned review and considering the bidirectional flow through doors, We carry out our study using 1.2 p/m/s for every door, meaning that a maximum number of 4.8 persons can pass through a 1-m width door per time slot (4 s). Staircase capacity is treated differently. Fruin [8] measures a maximum flow capacity of 5.5881 persons/s for a 5.6 width staircase, i.e., about 1 p/m/s. Weidmann [18] obtains 0.85 p/m/s for the same parameter. We stay at Fruin's estimate and use 1 p/m/s for each staircase in our example.

Cell capacity. The *pedestrian density* is the number of persons per square meter monitored at any time. This information is crucial for crowd safety and evacuation performance, as movements are dramatically reduced in highly dense areas. As density increases, pedestrian movements become constrained and flow rate consequently decreases. According to UK fire safety regulations, the maximum allowed density corresponds to 0.3 m^2 per standing person, a value that increases to 0.5 for public houses, to 0.8 for exhibition space, to 1.0 for dining places, to 2.0 for sport areas and to 6 for office areas. In our case study—gallery indoor space—the maximum capacity of each cell is calculated by assuming 0.8 m^2 per visitor, that is 1.25 persons per square meter.

5 Application

We next describe an application of the model of Sect. 3 to the safe evacuation of Palazzo Camponeschi, a building in L'Aquila (Italy) now and then used for exhibitions. Safety conditions of the building, which consists of 31 rooms including corridors (Fig. 2), are supervised by L'Aquila Fire Brigade and Civil Protection department. Rooms sizes vary in a large range, and consequently the average time of a person to cross them from door to door. The variegate building structure, as well as data on people attendance collected during events, made this study-case ideal for illustrating a general methodology for system sizing and development.

Cellular approximation of rooms. As explained in Sect. 4.1, we split each room in unit cells, each behaving as a (virtual) square room that can be traversed in a unit time slot. In practice, we embedded the building plan into a square grid as shown in Fig. 2. Red-marked cells are dummy nodes added to adjust pedestrian velocity on stairs (half that on flat surface, Sect. 4.2) to the unified monitoring time. To decide cell size, we look at both the error introduced by room approximation and the number of nodes in the resulting graph G. The latter is in an inverse proportion of cell size (bottom diagram in Fig. 3); the former varies irregularly with cell size (upper diagram of Fig. 3). We considered square cells up to 3.5×3.5 m (the short edge of the smallest room) and allowed no more than 500 nodes of G; then we selected the size that minimizes the largest error for all rooms. As shown in Fig. 3, 2×2 cells give the best approximation and imply 462 graph nodes. With 3.5×3.5 cells, the error rises a little bit over 40 (that is, twice that of 2×2) but G results in 116 nodes only. Summarizing, 3.5×3.5 leads to larger error but less CPU time; conversely, 2×2 causes larger CPU time but smaller error. We tested scenarios with both cell sizes in order to find the best efficiency/accuracy compromise.

IoT Infrastructure. To monitor and control the physical space, an IoT infrastructure consisting of sensing, computation, actuation and other elements is established. For the purpose of our experiment, we took advantage of CO_2 and temperature sensors to detect potential disaster, and of RFID tags/readers to gather the data necessary to feed the algorithm, as well as dynamically sensing/tracking crowd geolocation/movement in the physical environment [13, 14]. In our case, RFID tags are embedded into tickets that visitors should keep during their visit.

In normal situation, the sensors read CO_2 concentration, temperature and crowd geolocation frequently. This monitoring frequency, or its reciprocal that is the single cell crossing time, is derived from cells dimension and free flow speed: by the average free flow speeds of Sect. 4 we obtained time slots of 2.92 s for 3.5×3.5 cells, and of 1.67 s for 2×2 cells. The main goal of this application, to be run on a tablet, is to show a 2D-representation of the monitored space enriched with RFID-sensed contextual data that report where visitors are at any time located, and how they move in normal cases. In this mode, the optimal flow algorithm is periodically run to estimate the minimum evacuation time required under current conditions, a value that can be used to regulate visitor access so as to comply with safety conditions.

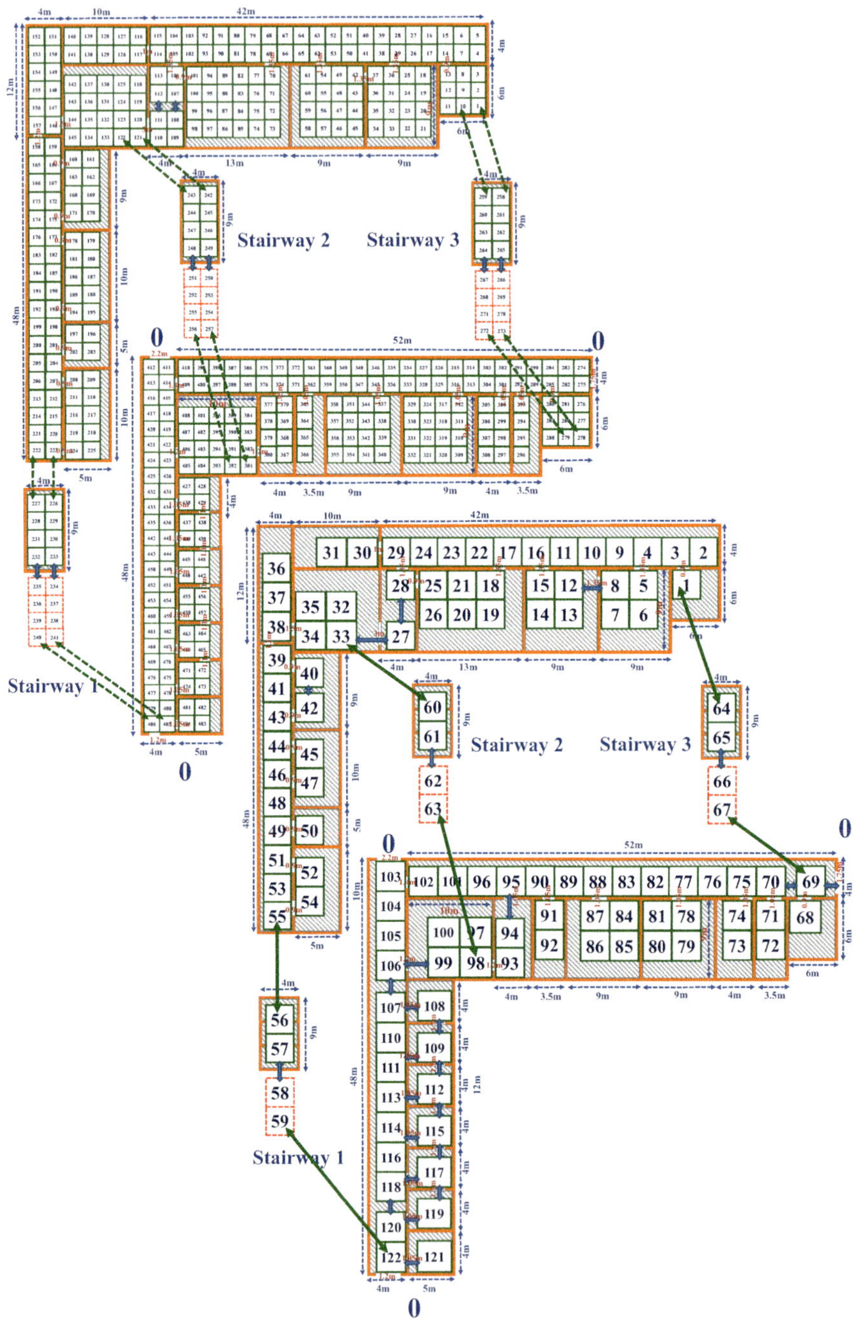

Fig. 2 Plan embedding of Palazzo Camponeschi into square grids with low (down-right) and high (up-left) resolution. For each resolution, ground floor down-right, first floor up-left. The area that is not covered by cells (error) is shown in gray

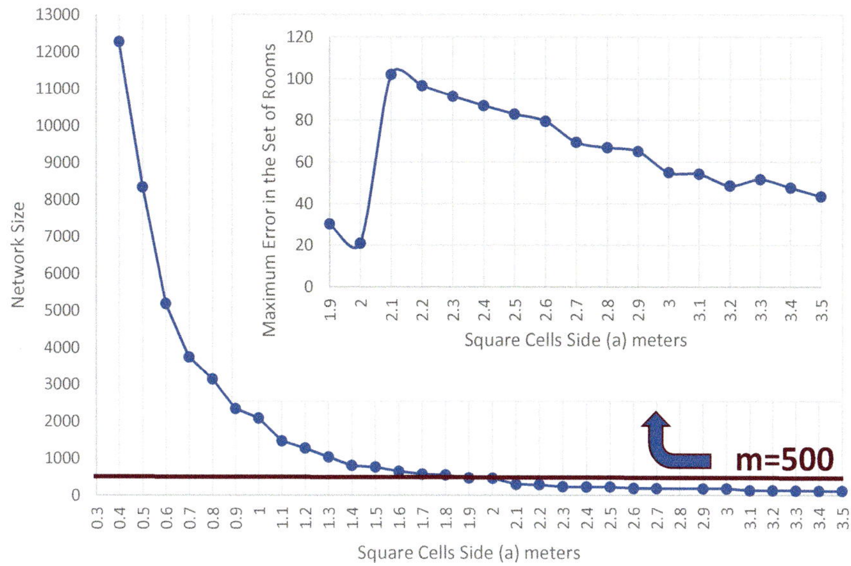

Fig. 3 Optimal cell size: maximum error (up right) and network size as a function of cell size

If a disaster detected, temperature/CO_2 detectors are read more frequently. In addition to map on dashboards, actuators of evacuation signs in each area show the best evacuation routes based on the discussed network model.

The network model. In the case here studied, plan embedding results in a graph with 116 (+6 dummy) nodes using 3.5 × 3.5 cells, or 462 (+24 dummy) nodes using 2 × 2 cells. Figure 4 corresponds to the lower resolution of Fig. 2 down-right, and includes node 0 as safe place. With 2 × 2 cells the network is very large and is not displayed. Adjacent cells are linked by arcs which allow flow inside the building. All arcs are assumed bidirectional except those towards the safe place. Time slots are as described above: 2.92 s for 3.5 × 3.5 cells and 1.67 s for 2 × 2 cells. Door capacities vary according to size and features. In our case, no more than 1.2 persons can pass through a 1-m arc (door or cell connection) per second.

Simulations. We next report the outcome of simulations using the optimization model under various conditions in terms of building occupancy, and compare it with static emergency maps based, as in usual circumstances, on shortest path route prescription. Simulations were first run for both cell sizes to evaluate the best choice. In all tests, we computed the minimum time required of N persons, randomly distributed in the building rooms, to reach a safe place. The code for simulation was written in OPL language and problems were solved by CPLEX version 12.8.0. We ran all the experiments on a Core i7 2.7 GHz computer with 16 Gb of RAM memory under Windows 10 pro 64-bits.

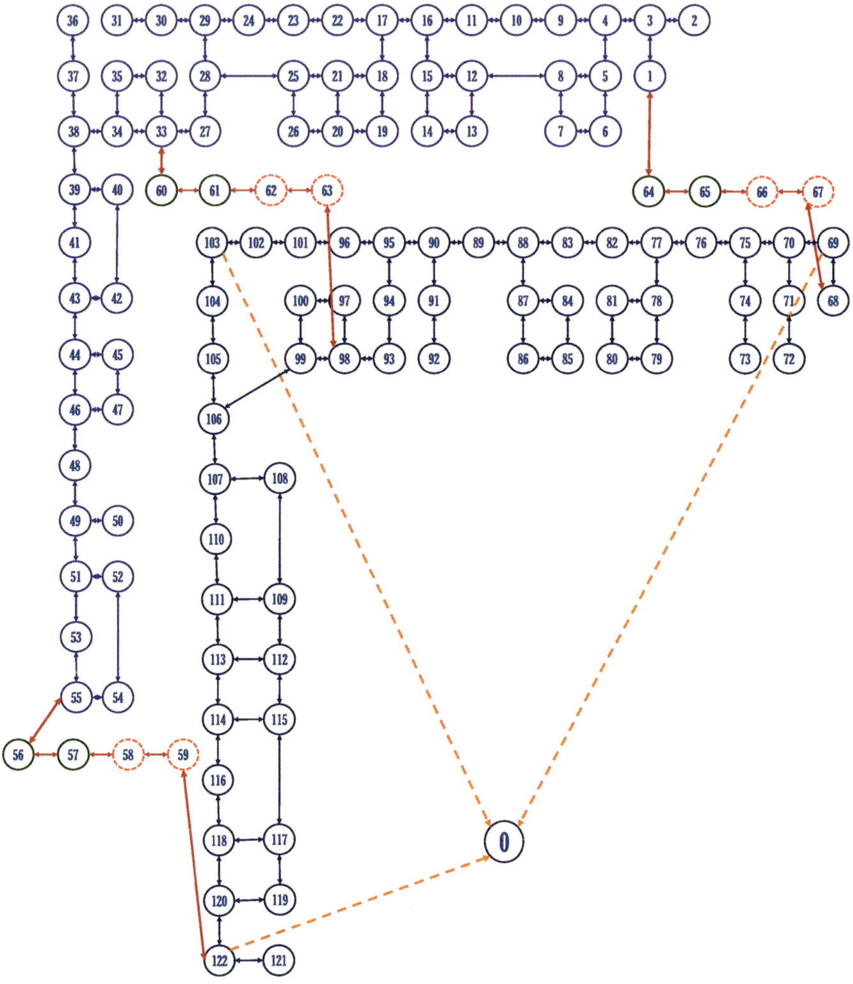

Fig. 4 Network associated with the plan of Fig. 2 down-right

Scenario 1. In the first simulation we suppose an initial occupancy of $N = 528$. This datum comes from an experiment performed in L'Aquila during the *Researchers' Night* event on 29 September 2017, when the simultaneous presence of 528 people in Palazzo Camponeschi was recorded as peak value. We solved problem (1)–(8) for $\tau = 1, 2, \ldots$ until a solution of value N is found. Table 2 reports the number of evacuees at each τ and the computation time of each solution step. Computation is done for both *low-* and *high-resolution* networks (respectively, 3.5×3.5 and 2×2 cells). With the low-resolution network, we get the evacuation and CPU times of Table 2 left: in terms of evacuation, everyone has reached a safe place in 1'25"; on the other hand, computation requires 2.89 s (presolve included) in the worst

Table 2 Evacuation and computation time—Scenario 1 (528): (*a*) 3.5 × 3.5 cells (time slots of 2.92 s); (*b*) 2 × 2 cells (time slots of 1.67 s)

$\tau(a)$	Evacuees (*a*)	CPU time (*a*) (s)	$\tau(b)$	Evacuees (*b*)	CPU time (*b*) (s)	$\tau(b)$	Evacuees (*b*)	CPU time (*b*) (s)
1	18	0.46	1	10	5.01	30	309	33.27
2	37	0.50	2	21	5.12	31	319	34.02
3	56	0.51	3	31	5.11	32	330	36.17
4	74	0.47	4	41	5.14	33	340	38.40
5	92	0.51	5	51	5.51	34	350	40.61
6	111	0.61	6	62	5.78	35	360	41.94
7	130	0.67	7	72	5.92	36	371	44.37
8	148	0.89	8	82	6.83	37	381	43.63
9	167	0.91	9	93	6.66	38	391	47.49
10	185	0.71	10	103	7.30	39	402	51.52
11	204	0.93	11	113	7.49	40	412	54.31
12	222	1.35	12	124	7.85	41	422	56.41
13	241	1.24	13	134	8.84	42	433	60.02
14	260	1.39	14	144	9.73	43	443	60.77
15	278	1.48	15	154	10.28	44	453	68.12
16	297	1.25	16	165	11.54	45	463	63.99
17	315	1.31	17	175	12.15	46	474	65.30
18	334	1.43	18	185	12.90	47	484	75.87
19	352	1.53	19	196	13.96	48	494	78.24
20	371	1.63	20	206	15.47	49	505	81.01
21	389	1.63	21	216	16.95	50	515	79.13
22	408	2.16	22	227	18.78	51	524	80.81
23	426	2.34	23	237	19.26	52	528	93.05
24	445	2.38	24	247	19.75			
25	463	2.53	25	257	21.73			
26	482	2.65	26	268	23.51			
27	501	2.76	27	278	27.08			
28	519	2.83	28	288	23.51			
29	528	2.89	29	299	30.35			

case, and is therefore totally compliant with real-time applications. Evacuation and CPU times for the high-resolution network are reported in Table 2 right. CPU time is now much larger (93.05 s) and the model appears inappropriate for real-time use, unless additional computational resource is deployed. We also see that everyone has reached a safe place in 1'27": hence we can conclude that sufficient accuracy is obtained using the low resolution network.

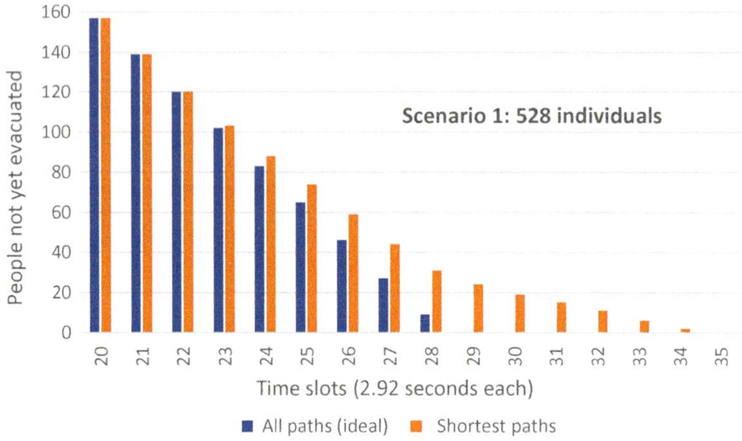

Fig. 5 *Ideal* versus *shortest paths* evacuation with 3.5 × 3.5 square cells: Scenario 1

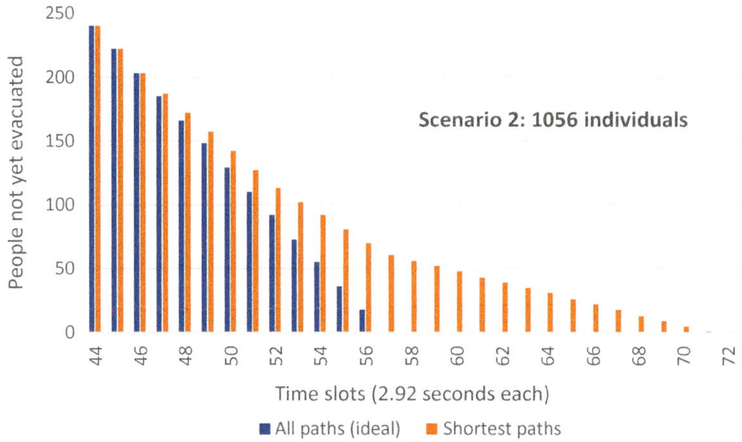

Fig. 6 *Ideal* versus *shortest paths* evacuation with 3.5 × 3.5 square cells: Scenarios 2

Ideal versus shortest paths. This first simulation depicts an ideal situation in which flows autonomously choose the best among all the available routes in the building. Of course, managing such an ideal evacuation is not easy and perhaps unpractical. As a general practice, in fact, evacuation is conducted through pre-determined routes. Considering this fact, then, we suppose that the prescribed evacuation routes are the shortest paths from any cell to the safe place. To evaluate this situation we find the subgraph of G formed by the shortest paths from any cell to 0 (as from static evacuation map), construct its time-indexed network and solve problem (1)–(4) for increasing τ. Evacuating 528 individuals takes of course more time: 1 min and 42". Thus, compared to the Netflow model we propose, shortest routes increase, in this case, the optimal evacuation time by 17% (Fig. 5).

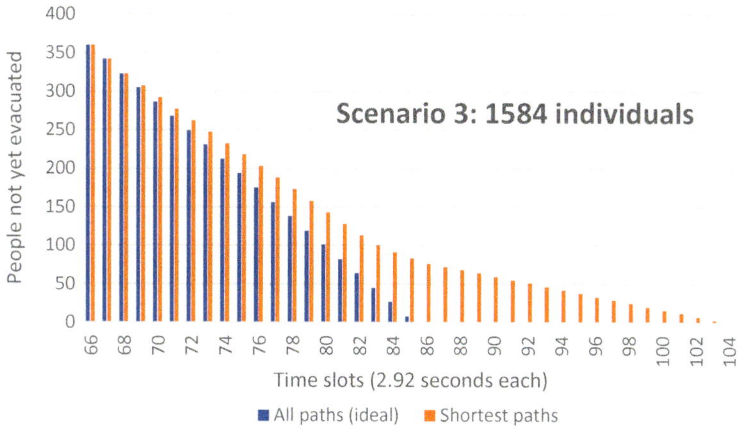

Fig. 7 *Ideal* versus *shortest paths* evacuation with 3.5×3.5 square cells: Scenarios 3

Scenario 2. In a second scenario, we repeated the simulation doubling the number of people in the building (1056 individuals). In this case everyone can reach a safe place after 57 time slots, i.e., 2 min and 47". In a second simulation, then, we again suppose that the prescribed evacuation routes are the shortest paths from any cell to the safe place. In this situation, evacuating 1056 individuals takes 3 min and 31". Shortest route evacuation increases in this case the optimum by over 26% (Fig. 6).

Scenario 3. In a third scenario, we repeated the simulation increasing the number of people to 1584 individuals, three times that of Scenario 1. In this case everyone can reach a safe place after 86 time slots, i.e., 4 min and 11". In a second simulation, then, we again suppose that the prescribed evacuation routes are the shortest paths from any cell to the safe place. In this situation, evacuating 1584 individuals takes 5 min and 4", over 19% more than optimal flows (Fig. 7).

In each of the three simulations, we observe plain flows for some time (1'04" in the first scenario, 2'14" in the second and 3'19" in the third). After that time, shortest routes start experiencing congestion, and evacuation is slowed down. The phenomenon is illustrated in the charts of the figures: as one can expect, the tail of people still in the building increases with initial occupancy.

We conclude with a short note on system usage. Leading people through the ideal evacuation paths needs proper actuation components. Optimal routes can be communicated to people using evacuation signs, lights, smart carpets, human agents, etc. Recent models, simulations and experiments gave evidence to a precious role played by individuals that, in case of emergency, emerge as crowd-leaders, see e.g. [2]. In the case study presented here, personnel positioned at almost each intersection was in charge of leading people to safe areas in case of emergency. A possible usage of our system would then envisage the periodical communication of best evacuation paths to personnel via a mobile device, which they are permanently equipped with.

6 Conclusions and Future Work

We use a run-time network flow model to support the rapid evacuation of people from a building in case of emergency. The model takes as input both static information (such as building dimension and structure, room capacities, door capacities) and run-time information (such as number of people in the building, number of people in specific rooms and corridors) acquired by IoT devices. The building topology is described by a graph. Run-time data are used to create a time-indexed acyclic digraph that models all the feasible transitions between adjacent rooms at any time. Minimizing evacuation time corresponds to solve max flow problems with non-linear capacities that model congestion. Preliminary evaluations have been conducted using data from a real case, and an ad-hoc IoT infrastructure has been designed.

The work presented opens ways to investigate a new model with cells occupied by one person at most. In this model, cell occupancy is regulated by individual reaction times, so that different types of individual behaviors can be described and supported. Possible constraints include grouping people (e.g., a couple or a family) into close paths: for instance, one can model specific individuals to be at any time within a prescribed distance. Further algorithmic research is however needed to assess the chances that such a model has to provide decision support in real-time.

More future work can include: (i) enriching the model with further design/run-time information to make plans more accurate; (ii) empirically evaluating the model in extensive scenarios, by comparing estimated versus real data (as, e.g., those generated during evacuation tests); (iii) completing the IoT infrastructure and app of Sect. 5 with the computational component here developed.

References

1. Abdelghany, A., Abdelghany, K., Mahmassani, H., Alhalabi, W.: Modeling framework for optimal evacuation of large-scale crowded pedestrian facilities. Eur. J. Oper. Res. **237**(3), 1105–1118 (2014)
2. Albi, G., Bongini, M., Cristiani, E., Kalise, D.: Invisible control of self-organizing agents leaving unknown environments. SIAM J. Appl. Math. **76**(4), 1683–1710 (2016)
3. Arbib, C., Muccini, H., Moghaddam, M.T.: Applying a network flow model to quick and safe evacuation of people from a building: a real case. In: Proceedings of the GEOSAFE Workshop on Robust Solutions for Fire Fighting, RSFF 2018, L'Aquila, Italy, July 19-20, 2018, pp. 50–61 (2018). http://ceur-ws.org/Vol-2146/paper02.pdf
4. Chen, P.H., Feng, F.: A fast flow control algorithm for real-time emergency evacuation in large indoor areas. Fire Saf. J. **44**(5), 732–740 (2009)
5. Choi, W., Hamacher, H.W., Tufekci, S.: Modeling of building evacuation problems by network flows with side constraints. Eur. J. Oper. Res. **35**(1), 98–110 (1988)
6. Daamen, W., Hoogendoorn, S.: Emergency door capacity: influence of door width, population composition and stress level. Fire Technol **48**(1), 55–71 (2012)
7. Fang, Z.M., Song, W.G., Li, Z.J., Tian, W., Lv, W., Ma, J., Xiao, X.: Experimental study on evacuation process in a stairwell of a high-rise building. Build. Environ **47**, 316–321 (2012)
8. Fruin, J.J.: Pedestrian Planning and Design, Metropolitan Association of Urban Designers and Environmental Planners. Inc, New York (1971)

9. Jiang, C., Deng, Y., Hu, C., Ding, H., Chow, W.K.: Crowding in platform staircases of a subway station in china during rush hours. Saf. Sci. **47**(7), 931–938 (2009)
10. Knoblauch, R., Pietrucha, M., Nitzburg, M.: Field studies of pedestrian walking speed and start-up time. Transp. Res. Rec. J. Transp. Res. Board **1538**, 27–38 (1996)
11. Kratchman, J.A.: An investigation on the effects of firefighter counterflow and human behavior in a six-story building evacuation. Ph.D. thesis, University of Maryland, College Park (2007)
12. Muccini, H., Arbib, C., Davidsson, P., Tourchi Moghaddam, M.: An iot software architecture for an evacuable building architecture. In: Proceedings of the 52nd Hawaii International Conference on System Sciences, pp. 678–687 (2019)
13. Muccini, H., Moghaddam, M.T.: Iot architectural styles. In: European Conference on Software Architecture, pp. 68–85. Springer (2018)
14. Muccini, H., Spalazzese, R., Moghaddam, M.T., Sharaf, M.: Self-adaptive iot architectures: an emergency handling case study. In: Proceedings of the 12th European Conference on Software Architecture: Companion Proceedings, p. 19. ACM (2018)
15. Patra, M., Sala, E., Ravishankar, K.: Evaluation of pedestrian flow characteristics across different facilities inside a railway station. Transp. Res. Procedia **25**, 4763–4770 (2017)
16. Schloter, M., Skutella, M.: Fast and memory-efficient algorithms for evacuation problems. In: Proceedings of the Twenty-Eighth Annual ACM-SIAM Symposium on Discrete Algorithms, pp. 821–840. SIAM (2017)
17. TranSafety, I.: Study compares older and younger pedestrian walking speeds. Road Manag. Eng. J. (1997)
18. Weidmann, U.: Transporttechnik der fußgänger: transporttechnische eigenschaften des fußgängerverkehrs, literaturauswertung. IVT Schriftenreihe **90** (1993)
19. Ye, J., Chen, X., Yang, C., Wu, J.: Walking behavior and pedestrian flow characteristics for different types of walking facilities. Transp. Res. Rec. J. Transp. Res. Board **2048**, 43–51 (2008)

From Pallets to Puppies: Using Insights from Logistics to Save Animals

M. Gentili, E. Gerber and K. Gue

Abstract The Kentucky Humane Society (KHS) is a private, nonprofit organization dedicated to saving every healthy, behaviorally sound animal that they take in. They are also the largest no-kill animal shelter in Kentucky, USA. Due to limited resources, the KHS has a great need for an optimized methodology for allocating available animals to their adoption facilities. Workers at the KHS believe that a capacity constraint is their most limiting factor. The purpose of this work is to develop an allocation model that will improve the throughput rate of the animals to be adopted, thereby freeing up extra capacity more quickly than current practices. The allocation model presented here assigns animals to adoption locations based on their expected length of stay at all available locations. The resulting allocation minimizes the expected length of stay for these animals. In order to make the model easy to use, we have implemented a software with a user-friendly interface.

Keywords Allocation model · User interface · Animal adoption

1 Background and Motivations

The Kentucky Humane Society (KHS) is a private, nonprofit organization located in Louisville, KY. Founded in 1884, it is the state's oldest animal welfare organization, and it is recognized as Kentucky's largest pet adoption agency and largest no-kill animal shelter. KHS is committed to saving every adoptable animal they take in. Thousands of animals are taken in throughout the year, with over 50% coming from overcrowded shelters and the remainder being owner surrenders and strays. The

M. Gentili (✉) · E. Gerber · K. Gue
Industrial Engineering Department, University of Louisville,
132 Eastern Parkway, Louisville, KY 40292, USA
e-mail: monica.gentili@louisville.edu

E. Gerber
e-mail: erin.gerber@louisville.edu

K. Gue
e-mail: kevin.gue@louisville.edu

M. Dell'Amico et al. (eds.), *A View of Operations Research Applications in Italy, 2018*,
AIRO Springer Series 2, https://doi.org/10.1007/978-3-030-25842-9_10

organization's success is measured by the number of lives saved, and KHS has successfully increased that number since 2003. In 2015 and 2016, their live release rate was 96%. This includes animals that were adopted, returned to owner, or (rarely) sent to rescues for specialized placement. Although KHS has had continued success, the organization wishes to better allocate adoptable animals to existing adoption facilities in order to free up space to accept more animals, and thus saving more lives. In 2012, KHS transitioned to admissions by appointment to ensure that they have adequate kennel space, eliminating euthanasia due to lack of space. This process has allowed more animals to be saved, but lack of space and inefficient utilization of space on the adoption floor are still among the most challenging issues that KHS has been facing.

Wanting to add a science-based approach, KHS asked researchers at the University of Louisville's Logistics and Distribution Institute, who normally deal with *pallets*, to apply their research to *puppies* with the goal of maximizing adoptions and decreasing length of stay.

The researchers analyzed shelter data; designed, implemented, and validated an optimization model; and implemented a software with a user-friendly interface to facilitate the use of the mathematical model by the Kentucky Humane Society staff. The optimization model assigns animals to adoption locations so that the overall expected length of stay (LOS) across all KHS facilities is minimized, without exceeding the holding capacity of any of the facilities.

The steps of the analysis are summarized in Sects. 2 and 3.

2 Project Phases

The goal of the project was to develop a mathematical model and deliver a software tool that can be used on a daily basis to better allocate animals to each adoption location and in order to minimize the expected length of stay. There are ten adoption locations (one intake location which is used as a surplus adoption site and nine adoption-only locations), and the optimization model was created to assign each animal to the location that would minimize the overall length of stay.

The ten adoption locations are as follows:

- Clarksville
- Dixie
- East Campus
- Fern Creek
- Hikes Point
- Main Campus (intake and surplus)
- Pewee Valley
- Preston
- Springhurst
- St. Matthews

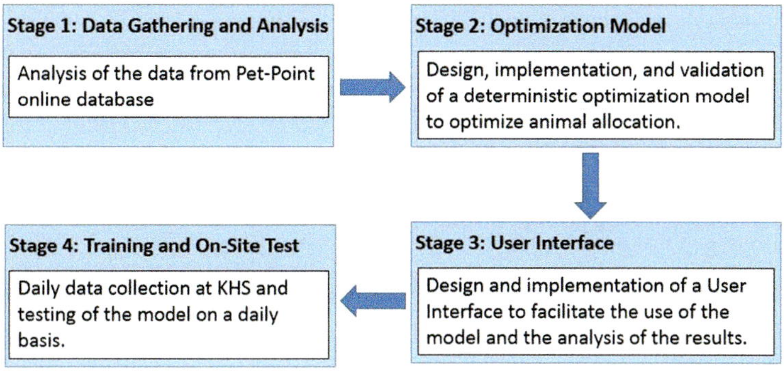

Fig. 1 Project phases

Each location has a set amount of spaces (of fixed sizes) to house the animals, as well as a certain clientele that it caters to. By implementing an optimization model that takes into account which types of animals are adopted most quickly at each location, as well as space availability, the animals are more likely to be adopted at a faster rate, allowing space to open up to accommodate more incoming animals.

The project was carried out in four stages which are summarized in Fig. 1 and detailed in the rest of this paper.

3 Methodology

Stage 1: Data Gathering and Analysis

KHS uses an online data management system, PetPoint, to track all animals within their system. PetPoint is a comprehensive and user-friendly online database. There are more than 60 different types of reports that can be pulled from PetPoint. The database holds all data since KHS began using it in 2009. For our allocation model we used data from April 2014 to March 2016, and we only considered dogs. The analysis of the data focused on estimating the expected length of stay for different dogs at different locations. Adoption rate at each location is based on several factors established by KHS (breed, size, age, health, behavior, and appearance). However, given the available data, only two of these characteristics were used to generate statistically significant estimates: size and age. Animals were divided into 10 different groups according to their size (Puppy, Small, Medium, Large/Extra Large) and their age (0–6 months, 7 months–3 years, 3–7 years, 7+ years). Data was collected and utilized to calculate a probability matrix for animals within each group to be adopted in a given location. Probabilities for various lengths were calculated using the corresponding probability matrix. An expected value using each

Table 1 LOS estimates per location (days)

Location	Puppy	Small			Medium			Large		
	0–6m	7m–3y	3–7y	7+y	7m–3y	3–7y	7+y	7m–3y	3–7y	7+y
Clarksville	6.6	9.8	10.3	12.0	9.4	11.0	11.1	7.4	9.5	12.5
Dixie	7.8	10.1	9.6	11.8	9.5	11.5	10.2	10.7	12.5	12.5
East Campus	7.4	8.3	11.5	11.2	10.0	11.1	12.5	10.1	11.9	10.9
Fern Creek	6.5	8.2	10.3	11.0	9.8	9.6	12.5	9.5	9.3	12.5
Hikes Point	7.5	9.7	9.5	8.7	9.4	8.4	11.2	9.8	10.8	12.5
Main Campus	6.9	8.8	9.1	10.2	8.7	10.2	10.9	8.9	10.3	11.1
Pewee Valley	4.2	4.2	0.0	0.0	12.5	0.0	0.0	6.8	0.0	0.0
Preston	6.8	9.5	10.8	11.3	8.7	12.0	12.2	12.5	12.5	0.0
Springhurst	6.9	8.9	8.9	6.5	9.6	9.7	0.0	10.4	9.0	10.2
St. Matthew	6.8	6.6	8.0	9.2	8.5	9.1	9.5	8.4	8.6	11.2

probability density function was calculated and utilized for the model. The resulting current LOS estimates are reported in Table 1. As it can be noticed from the table, there are some differences from one location to another. This was not surprising to the workers at KHS since demand for adoption is highly correlated with demographic characteristics of residents in different geographic areas.

Stage 2: Design, Implementation, and Validation of the Optimization Model

The mathematical model is a min-sum linear assignment model [1, 3] designed to allocate animals to locations each day so as to minimize the total expected length of stay across the system, while respecting constraints on available cages of varying size. The model input includes the following data:

- estimates of the expected LOS (in days) for each animal group at each location;
- number of animals in each group to be allocated;
- available capacity for each cage size at each location.

The model output includes:

- allocations of animals to locations such that:
  - total expected LOS is minimized;
  - all animals are allocated respecting the available capacity constraints.

There are a few assumptions associated with the allocation model. First, KHS has one intake facility and 9 offsite adoption centers located throughout Louisville Metro. Second, the allocation model only considers animals that are ready to be allocated on a given day. Third, the model assumes that animals will not be transferred between locations once they have already been allocated. Finally, it is assumed that dogs are able to go into their appropriate cage size or one size larger. That is:

- cages for puppies can allocate only puppies (group 1);
- cages for small dogs can allocate small dogs (groups 2, 3 and 4) and puppies (group 1);
- cages for medium size dogs can allocate medium size dogs (groups 5, 6, and 7) and small dogs (groups 2, 3 and 4);
- cages for large dogs can allocate large dogs (groups 8, 9, and 10) and medium size dogs (groups 5, 6, and 7).

This was found to be common practice at KHS currently, and it allows the model to explore additional options in order to improve the overall assignment.

The mathematical formulation of the model follows.

$$\min \sum_{j \in J} \sum_{k \in K} x_{jk} E[LOS_{jk}] \tag{1}$$

$$\sum_{j \in J} x_{jk} = n_k \quad \forall k \in K \tag{2}$$

$$\sum_{k \in K(s)} x_{jk} \leq c_{js} \quad \forall j \in J, \quad \forall s \in S \tag{3}$$

$$x_{jk} \geq 0 \quad integer \quad \forall j \in J, \quad \forall k \in K \tag{4}$$

where:

- $k \in K$: index set for animal group;
- $j \in J$: index set for adoption location;
- $s \in S$: index set for animal size;
- n_k: number of animals in group k to be assigned on a given day;
- c_{js}: capacity at location j for animals of size s;
- $E[LOS_{jk}]$: expected length of stay (in number of days) of an animal in group k in location j;
- x_{jk}: number of animals in group k allocated to location j;
- $K(s) \subseteq K$: groups of animals which can be assigned to a cage for animals of size s.

The objective function minimizes the cumulative expected length of stay for all animals allocated to all locations. Constraints (2) ensure all available animals are allocated. Constraints (3) ensure the allocation does not exceed capacity at any given adoption location.

There are several efficient and specialized algorithms that can be used to solve this model, but due to the size of the real problem, a general purpose solver for integer models is enough. Hence, the model was implemented using Microsoft Excel and solved using the Open Solver Excel VBA add-in [2, 4].

To validate the model we compared the results from the optimized allocation with the results from the manual allocation performed by KHS for the period August

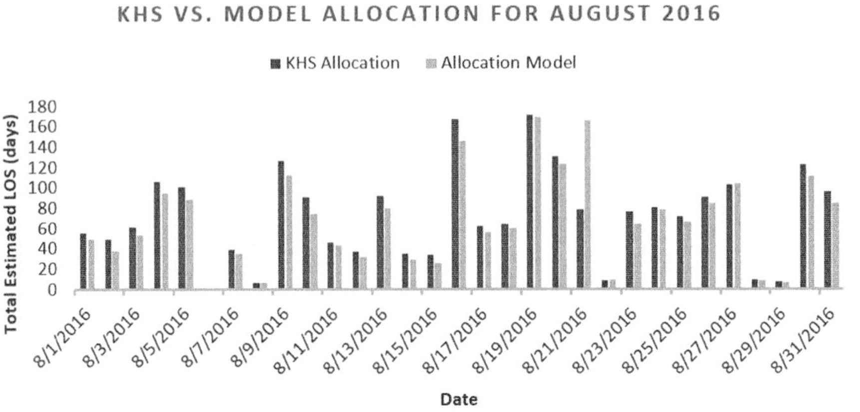

Fig. 2 August overview for minimizing LOS for dogs

2016–September 2016. Figure 2 shows the comparison between the total length of stay associated with the optimized allocation versus the total length of stay associated with the manual allocation for the month of August 2016. Similar results were found during September 2016. As expected, our allocation model obtains a lower LOS than the manual solution on most of the days. The comparison revealed a few exceptions (August 21, 2016 and August 22, 2016) due to the fact that the manual allocation allowed a solution which was infeasible for the model. For example, on August 21, 2016, KHS decided to place an animal in a cage two sizes larger at a location that had a much lower LOS, while the model only allows an animal to go into its appropriate size cage or one size cage above. This was also the case for the other dates where the manual solution resulted in a shorter LOS than the optimized solution. Since this is not common practice but it was due to the current capacity status of the adoption location and to the particular type of animals, this allocation rule was not implemented in the model.

The overall LOS (in days) for dogs during August 2016 and September 2016 was 4444.7 for the KHS allocation and 4069 for the model allocation, resulting in an overall decrease of 10.9% when using the allocation model. The average, minimum, and maximum LOS values (in days) for the two allocation methods during August 2016 and September 2016 were 75.3, 6.6, and 185 for KHS and 66.7, 4.2, and 169 for the model allocation, respectively. As expected, the allocation model resulted in smaller values for the average LOS, minimum LOS, and maximum LOS. If KHS had used the allocation model during this time period and all the animals had been adopted, they would have freed up on average a total of 375.1 days of cage space. This would have given them the ability to adopt an estimated 20.2 additional dogs during this two-month period.

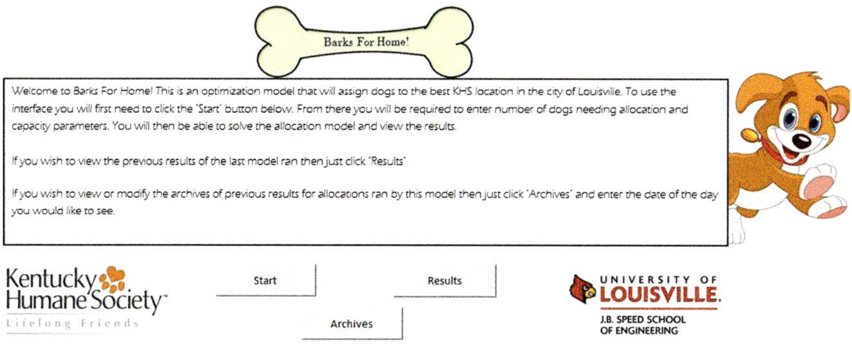

Fig. 3 Welcome spreadsheet of the application

Stage 3: User Interface Implementation

This stage of the project focused on the development of a user interface to facilitate the use of the mathematical model by the Kentucky Humane Society staff. The interface was developed in Visual Basic for Applications for Microsoft Excel. The Open-Solver Excel was used and installed into the program to enhance its allocation capabilities. The software allows the users not only to run the optimization model daily but also to store the daily solution and perform analysis of the results over time. Figure 3 shows the welcome spreadsheet of the application containing a brief description of what the program does along with simple starting instructions.

Stage 4: Documentation, Training, and On-site Model Testing

The last stage of the project was devoted to (i) creating detailed instructions for using and maintaining the model; (ii) conducting training to hand off the finished model to KHS users; and (iii) testing the model by using real data for at least a six month period. KHS has been using the model to manage its daily allocation decisions since the beginning of 2017.

4 Conclusions and Next Steps

The initial results show that the allocation model provides a substantial improvement compared to the KHS allocation. This outcome not only verified that the allocation model worked properly, but it also assured that this will allow KHS to make more efficient use of the limited space they have. The allocation model created for the KHS is a good start in improving their adoption process. However, the employees at KHS are still the experts in this field. Thus, the allocation model should only be used as a guide in the allocation process. There are exceptions that may be made by the employees, utilizing rules not included in the model that would improve the final allocation. Overall, employees at KHS were satisfied with the developed tool which makes their decision making process faster and more effective.

Follow-up projects include enhancement to the current allocation model into two directions. The solution of the model can be significantly improved by incorporating the preparation period before an animal can be ready for adoption. Indeed, because the number of dogs to be sent out for allocation is known not only for today but also for the next several days, model performance can be improved by taking into account the dogs that will become available in the near future. Additionally, the randomness of the length of stay (LOS_{jk}) parameter in the model affects the available capacity at each location and therefore the overall allocation process. Hence, accounting for the randomness of this parameter in the model would improve the overall allocation.

Acknowledgements The authors would like to thank Robin Vincent from the Kentucky Humane Society for all her support and extensive knowledge throughout this research project. The authors are also grateful to the two students who worked on the project: Stefanie Olga who helped with the activities of Stage 1 of the project, and Mark Cummins who implemented the software.

References

1. Burkard, R., Dell'Amico, M., Martello, S.: Assignment Problems, vol. 106. Siam, Philadelphia (2012). Revised reprint
2. Mason, A.J.: Opensolver-an open source add-in to solve linear and integer progammes in excel. In: Operations research proceedings 2011, pp. 401–406. Springer (2012)
3. Munkres, J.: Algorithms for the assignment and transportation problems. J. Soc. Ind. Appl. Math. **5**(1), 32–38 (1957)
4. OpenSolver: https://opensolver.org/

Improving Social Assistance Services for Minors and Disabled People by Using Multiobjective Programming

Lorenzo Lampariello, Andrea Manno and Simone Sagratella

Abstract Relying on multiobjective programming techniques, we have developed an optimization software to improve the services provided by the social co-operative OMNIA. OMNIA's mission is to supply home-care assistance to children and people with disabilities. Our method is intended to optimize the social workers' shift planning aiming at, on the one hand, maximizing the overall quality of the social care services, on the other hand, minimizing costs associated with OMNIA's activity. In particular, our software provides Pareto optima of the resulting (difficult) bi-objective model by resorting to both a standard a priori weighted-sum approach and a new MINLP no-preference (hypervolume maximization-type) method. The product of this research is successfully employed by OMNIA.

Keywords Multiobjective programming · Nonlinear programming · Integer programming

1 OMNIA Social Co-operative

The OMNIA social co-operative (http://www.cooperativaomnia.com/public/) provides social, health and education home care services and school assistance to people with disabilities or children in poor living conditions. In particular, these services can be categorized in: material assistance, consisting in personal hygiene and meal aid,

---

L. Lampariello
Department of Business Studies, Roma Tre University, Via D'Amico 77, 00145 Rome, Italy
e-mail: lorenzo.lampariello@uniroma3.it

A. Manno (✉)
Dipartimento di Elettronica, Informazione e Bioingegneria, Politecnico di Milano, Via Ponzio 34/5, 20133 Milan, Italy
e-mail: andrea.manno@polimi.it

S. Sagratella
Department of Computer, Control, and Management Engineering Antonio Ruberti at Sapienza University of Rome, Via Ariosto 25, 00185 Rome, Italy
e-mail: sagratella@diag.uniroma1.it

M. Dell'Amico et al. (eds.), *A View of Operations Research Applications in Italy, 2018*, AIRO Springer Series 2, https://doi.org/10.1007/978-3-030-25842-9_11

support in health services and care of home environment; education assistance, consisting in enhancement/maintenance of cognitive skills, incentives for social autonomy and promotion of motor, recreational and laboratory activities; school assistance for minors. OMNIA can rely on a team of social health workers and educators.

The services provided by OMNIA have a great social impact since they are addressed to people with very severe disabilities or children in extremely difficult conditions, and they are operated in a complicated territory with isolated municipalities which are difficult to reach. Therefore, organizing the assistance services offered by OMNIA is a very complex task. Furthermore, it would be economically unsustainable for the co-operative to hire a private consultant to efficiently address this issue. For these reasons, this work is intended to provide OMNIA with an easy-to-use software to optimally manage the most critical organizational aspect: the social workers' shifts planning. The software is based on quantitative methods typical of Operations Research: it is able to optimize different performance criteria with the aim of maximizing the collective well-being intended as

- well-being of the service recipients
- better working conditions for the social workers
- easing the organizational work of the OMNIA staff.

More specifically, one can identify 14 different types of assistance services. It should be noted that, although both categories of social workers can, in principle, perform any type of service, some tasks require the exclusive (or at least preferable) assistance of a specific type of social worker. Each recipient requires a *provision of assistance* which, in general, is made up of a set of services (belonging to the aforementioned 14 categories), and for which it is specified the delivery mode (time duration, time slot, frequency, place, etc.). Different assistance provisions, if consistent by nature and schedule, can in turn be merged a priori into aggregated packages, namely *jobs*. OMNIA commits the provision of assistance to the available social workers. In turn, the social workers are remunerated for their services by OMNIA through the fundings that the government entrusts to OMNIA for its activity. OMNIA has to plan and manage the assistance shifts guaranteed by the social workers. To this end, the recipients indicate days and time slots (with an accuracy of 30 min) in which they need assistance, while the social workers declare the time slots (morning, afternoon or both) in which they can provide the service. Note that they are, however, ready to work every day of the week; furthermore, social workers sustain the transportation costs to join the place where their service is required. Finally, every social worker has to perform his work exclusively in the time slot predetermined for the service shift. Taking into account the aforementioned constraints, and the number of social workers and recipients involved, the organization of the assistance services over the time horizon turns out to be a very complicated task. Without the aid of appropriate support tools, this activity can require a considerable consumption of resources and time. At the same time, optimal shifts planning is a fundamental requirement to guarantee regularity and punctuality of the assistance services and ensure the best possible working conditions for the social workers, so as to offer a high quality service.

The introduction of an analytical tool for the shifts planning can improve the global level of service, strongly affecting both the previous criteria. Furthermore, it allows to accelerate and lighten the organizational burden of the OMNIA staff, with further positive implications on the quality of the offered services.

On the basis of these considerations, a software has been developed, which exploits optimization techniques for the optimal organization of the assistance services. First, the identification of the optimal shifts planning has been formulated as a bi-objective constrained mathematical program; then, suitable algorithms for the efficient solution of the resulting problem have been devised and implemented in a user-friendly software.

For the technical details of these aspects, the reader is referred to the following Sect. 2. The main elements characterizing the proposed approach are listed below.

- Given a social worker i and a job j, a measure q_{ij} has been devised to quantify how much the social worker i is able to execute the assistance services included in job j. By disaggregating a job in the single component assistance provisions, and in turn each assistance provision in the single types of component services, it is possible to associate to each couple "social worker-type of service" a compatibility score. The score is equal to 1 if the worker is able to execute the specific type of service, and 0 otherwise. The overall coupling measure associated to "worker i-assistance provision" is given by the mean of the scores obtained by the worker i over all the types of services included in the assistance provision. Finally, if the job j is composed by multiple assistance provisions, the overall measure q_{ij} of the pair "worker i-job j" is given by the sum of the quality values for the single assistance provisions belonging to job j.
- To each pair "worker i-job j" is associated a transportation cost c_{ij} representing the distance (in km) traveled by worker i to execute job j and to return to his municipality of residence.
- In accordance with the OMNIA staff, days are divided into the morning and the afternoon intervals. The availability of workers and recipients in each time interval is managed in a sort of pre-processing; as a result, only combinations meeting the needs in terms of schedules will be considered.
- The following objectives are considered:

(i) maximization of a measure of the overall "quality" for the assignment of the workers to the jobs to be executed;
(ii) minimization of the transportation costs sustained by the workers to execute the jobs.

With objective (i) we want to exploit as much as possible the skills of the workers in order to offer a high quality service to the recipients; objective (ii) is aimed at simplifying the work of the social workers: getting easily to the workplace has evident positive impacts on the punctuality and on the actual duration of the offered services, on the mood of the workers and on the costs of the service itself. Although in different way, both objectives contribute to improve the overall quality of the care services and, ultimately, the social well-being.

- The provided software implements two different (bi-objective) optimization strategies for the management of the assistance services. A first strategy allows the user to weight the two objectives so that it is possible to orientate the solution towards specific needs. A second strategy autonomously produces the Pareto efficient solution that optimizes a certain measure of global performance, to be described later.
- The dimensions of the problem are quite considerable. OMNIA activity involves 25 social workers, 72 recipients and a total of 170 assistance provisions. These data are only partial.

2 The Mathematical Programming Formulation

In light of the objectives above, OMNIA has to identify the most appropriate social workers to provide the recipients with the assistance services, in a time horizon of 6 days a week, for every morning and every afternoon. For the sake of simplicity and without loss of generality, from now on we will refer to the optimization problem corresponding to a single half-day (morning or afternoon).

The set of jobs provided by the social co-operative to the recipients is denoted by L, while O denotes the set of social workers. Each worker is in charge to execute (at most) one of the jobs belonging to the set L. Clearly, each job is required to be fully completed. Overall, the problem consists in assigning each worker to (at most) a job so as to minimize the overall transportation costs and, at the same time, maximize the overall "quality" of the services provided by the social co-operative.

The bi-objective mathematical model of the problem is as follows:

$$\begin{array}{ll} \underset{x}{\text{minimize}} & (f_1(x),\ f_2(x))^T \\ \text{s.t.} & x \in X \\ & x \in \{0,\ 1\}^{|O||L|}, \end{array} \tag{1}$$

where $x \triangleq [x_{ij}]_{i \in O,\, j \in L}$, and $|O|$ and $|L|$ denote the cardinality of sets O and L, respectively; clearly, $x_{ij} = 1$ if worker i is assigned to job j, and $x_{ij} = 0$ otherwise. The objectives $f_1(x) \triangleq \sum_{i \in O,\, j \in L} c_{ij} x_{ij}$ and $f_2(x) \triangleq -\sum_{i \in O,\, j \in L} q_{ij} x_{ij}$ suitable take into account the transportation costs and the quality of the assignments, respectively. The feasible region $X \triangleq \{x \in \mathbb{R}^{|O||L|} : \sum_{i \in O} x_{ij} = 1\ \forall j \in L,\ \sum_{j \in L} x_{ij} \leq 1\ \forall i \in O\}$ is made up of two family of standard constraints. The former imposes that each job $j \in L$ must be executed by exactly one worker $i \in O$. The latter entails that each worker $i \in O$ executes at most one job $j \in L$.

Solving multi objective problems consists in finding the best trade-off between the conflicting goals according to an order relation, as the classical Pareto dominance relation (see [12]): w.r.t. a Pareto optimum, one cannot think to improve on an objective without deteriorating the remaining one. More precisely, a feasible solution $\overline{x}$ to the bi-objective problem (1) is said to be Pareto optimal if there not exists any feasible solution x such that either $f_1(x) \leq f_1(\overline{x})$ and $f_2(x) < f_2(\overline{x})$, or $f_1(x) < f_1(\overline{x})$ and $f_2(x) \leq f_2(\overline{x})$.

3 Solution Methods

Problem (1) is, in general, difficult to solve. The bi-objective nature of the model, as well as the integrality constraints make complicated the design of efficient solution methods. Taking into account these critical aspects, relying on suitable scalarization techniques, the bi-objective problem (1) can be reconducted to an appropriate optimization problem with a single-objective (and integrality constraints). To perform this tansformation, different strategies can be used: considering the needs of OMNIA, and following their specific requests, in this work we refer to the paradigm of a priori approaches and to that of methods without preferences (see [12]).

3.1 Maximization of the Hypervolume: A Method Without Preferences

The aim of methods without preferences is to determine, among all the Pareto solution, the one with specific and distinctive features. Considering the bi-objective problem (1), let $f_1^{\max}$ and $f_2^{\max}$ be the "worst" values that the objectives can achieve if evaluated individually on the feasible set. Then, each feasible point $\widetilde{x}$ identifies in the objectives space a rectangle defined by the points $(f_1^{\max}, f_2^{\max})$ and $(f_1(\widetilde{x}), f_2(\widetilde{x}))$, with basis $f_1^{\max} - f_1(\widetilde{x}) \geq 0$ and height $f_2^{\max} - f_2(\widetilde{x}) \geq 0$: clearly, the feasible points with corresponding objectives values belonging to the above rectangle are all dominated by $\widetilde{x}$. Referring to the *hypervolume* technique (see e.g. [1, 10, 13]) and in the spirit of [11], in the light of the previous observations, the very natural idea arises to seek a feasible point that maximizes the area of the corresponding rectangle:

$$\begin{array}{ll} \underset{x}{\text{maximize}} & h(x) \triangleq (f_1^{\max} - f_1(x))(f_2^{\max} - f_2(x)) \\ \text{s.t.} & x \in X \\ & x \in \{0, 1\}^{|O||L|}. \end{array} \tag{2}$$

Proposition 1, showing that by addressing (2) one can recover a Pareto optimum for (1), provides the fundamental link between the single-objective problem (2) and the original bi-objective problem (1). As a consequence, this result lays down the theoretical foundation on which our scalarization technique builds upon.

Proposition 1 *Any global optimum $\overline{x}$ for problem (2) such that $(f_1^{\max} - f_1(\overline{x}))(f_2^{\max} - f_2(\overline{x})) > 0$ is a global Pareto optimum for problem (1).*

Proof Suppose by contradiction that there exists $\widetilde{x} \in X \cap \{0, 1\}^{|O||L|}$ such that, without loss of generality, $f_1(\widetilde{x}) < f_1(\overline{x})$, $f_2(\widetilde{x}) \leq f_2(\overline{x})$. In turn, $(f_1^{\max} - f_1(\widetilde{x})) > (f_1^{\max} - f_1(\overline{x})) > 0$, and $(f_2^{\max} - f_2(\widetilde{x})) \geq (f_2^{\max} - f_2(\overline{x})) > 0$ and, hence, $(f_1^{\max} - f_1(\widetilde{x}))(f_2^{\max} - f_2(\widetilde{x})) > (f_1^{\max} - f_1(\overline{x}))(f_2^{\max} - f_2(\overline{x}))$. But this, in turn, is in contradiction with the optimality of $\overline{x}$ for (2).

Proposition 2 characterizes, among all the Pareto optima for problem (1), the one given by (2). We point out that clearly, in the light of the very nature of a priori methods, it is rather sensible to show that the computed Pareto optimum is equipped with significant properties and distinctive features, as detailed next.

Proposition 2 *Any solution* $\overline{x}$ *of problem (2) such that* $(f_1^{\max} - f_1(\overline{x}))(f_2^{\max} - f_2(\overline{x})) > 0$ *enjoys the following properties:*

(a) if a feasible point $\widetilde{x} \in X$ *is such that* $f_1^{\max} - f_1(\widetilde{x}) = \alpha\big(f_1^{\max} - f_1(\overline{x})\big)$ *for some* $\alpha \geq 1$*, then* $\alpha\big(f_2^{\max} - f_2(\widetilde{x})\big) \leq f_2^{\max} - f_2(\overline{x})$*;*
(b) if a feasible point $\widetilde{x} \in X$ *is such that* $f_2^{\max} - f_2(\widetilde{x}) = \beta\big(f_2^{\max} - f_2(\overline{x})\big)$ *for some* $\beta \geq 1$*, then* $\beta\big(f_1^{\max} - f_1(\widetilde{x})\big) \leq f_1^{\max} - f_1(\overline{x})$*.*

Proof The claims are due to the following relations:

$$\alpha = \frac{f_1^{\max} - f_1(\widetilde{x})}{f_1^{\max} - f_1(\overline{x})} \leq \frac{f_2^{\max} - f_2(\overline{x})}{f_2^{\max} - f_2(\widetilde{x})} = \frac{1}{\beta},$$

where the inequality holds in view of the optimality of $\overline{x}$ for (2).

Roughly speaking, in view of Proposition 2, the scalarization technique proposed here takes into account the objectives in a balanced way, as also witnessed by the results of the practical implementation of the management software devised for OMNIA (see Sect. 4). Note that the results above, that have been studied in a different context in the recent paper [9], actually hold for general functions f_1 and f_2, since they are consequences of the very concept of optimality.

Let, from now on, $\overline{x}$ with $(f_1^{\max} - f_1(\overline{x}))(f_2^{\max} - f_2(\overline{x})) > 0$ be a global optimal point for problem (2). In a nutshell, addressing (2) has the following prominent advantages and distinctive features that are enforced by the approach itself:

(i) if one decides to choose a feasible solution different from $\overline{x}$ that entails an improvement in terms of either f_1 or f_2 w.r.t. $\overline{x}$ *by a given factor*, then the price to pay is the corresponding worsening of the remaining objective *by at least the same factor*. Thus, choosing any other feasible solution improving on $\overline{x}$ w.r.t. transportations costs or quality of the overall assignment, implies the willingness to accept a worsening of the other objective *at least as large* (see Proposition 2);
(ii) the solution $\overline{x}$, maximizing the area of the induced rectangle, dominates the largest amount of points that dominate the worst point. Therefore, in a sense, $\overline{x}$ dominates the most the worst point;
(iii) finding a solution of (2) does not depend on the units of performance measurements of the objectives, thus allowing to "robustly" deal with non homogeneous objectives such as transport costs and quality of service measures like in the application here considered.

In summary, Propositions 1 and 2, along with the related properties (i)–(iii), justify the soundness of the approach: in fact they show that addressing the single-objective problem provides one with a peculiar and significant solution for the original bi-objective program (1). On the other hand, of course, passing from (1) to (2), which enjoys the remarkable features (i)–(iii), comes at a price: in fact, in general, computing a solution of a nonlinear mixed integer optimization problem (MINLP) as (2) is a difficult task. Note also that, even if f_1 and f_2 are linear, the objective function in (2) is in general nonconvex. Thus, the question about how one can practically solve (2) still remains open. For these reasons, and in the light of Proposition 1, it is convenient to introduce the perturbed problem

$$\begin{array}{ll} \underset{x}{\text{maximize}} & h_\varepsilon(x) \triangleq (f_1^{\max} + \varepsilon - f_1(x))(f_2^{\max} + \varepsilon - f_2(x)) \\ \text{s.t.} & x \in X \\ & x \in \{0, 1\}^{|O||L|}, \end{array} \tag{3}$$

where $\varepsilon > 0$ is a fixed parameter. Clearly, with $\varepsilon = 0$, problem (3) reduces to problem (2). We resort to the perturbation ε in order to get strictly positive terms in the product defining the single-objective function: $f_1^{\max} + \varepsilon - f_1(x)$, $f_2^{\max} + \varepsilon - f_2(x) > 0$, for every $\varepsilon > 0$ and for any x belonging to an open superset of X. This key property, on the one hand, following the line of reasoning in Proposition 1, allows one to prove that any global solution of (3) is a global Pareto optimum of the original problem (1) that enjoys the nice properties (i)–(iii) described above. On the other hand and more importantly, makes it possible to recover useful generalized convexity properties, as detailed next.

Proposition 3 *Function h_ε, for every $\varepsilon > 0$, is pseudoconcave on an open convex set containing X.*

Proof Function h_ε is semistrictly quasiconcave, thanks to the convexity of f_1 and f_2 (see [2, Corollary 5.18]). The claim follows from [2, Theorem 5.17(a)], since f_1 and f_2 are continuously differentiable and $f_1^{\max} + \varepsilon - f_1(x)$, $f_2^{\max} + \varepsilon - f_2(x) > 0$ for every $\varepsilon > 0$ and for any x belonging to an open superset of X.

Note that the claim in Proposition 3 holds more in general for convex and continuously differentiable functions f_1 and f_2. In view of its very definition, it is precisely the pseudoconcavity that makes a stationary point of h_ε on X a global optimum (see e.g. [3]). This result is crucial for the viability of the proposed approach. In fact, the practical solution of problem (3) can be obtained by means of branch-and-bound-type algorithms in a MINLP framework (see [4]) and so, in the end, by the iterative solution of nonlinear continuous (single-objective) relaxations through classical Newton type methods. In turn, the use of these methods is made possible by the sufficiently concave nature of h_ε over X.

We remark that, in the context of the application addressed here where f_1 and f_2 are linear, problem (3) reduces to a classical Quadratic Assignment Problem (QAP). Therefore, alternatively, one could also resort to any numerical method for QAPs [6–8].

Summing up, a careful tuning of the model allows one to devise an innovative approach that combines nonlinear optimization techniques with the appealing strategy of maximizing the hypervolume as a single criterion to evaluate conflicting criteria to be optimized simultaneously.

3.2 *The Weighted-Sum Method*

To take into account the subjective preferences of the user (in terms of either the first or the second objective), it is possible to use the standard weighted-sum method (see [12]). In this case we have

$$\begin{array}{ll} \underset{x}{\text{minimize}} & w_1 f_1(x) + w_2 f_2(x) \\ \text{s.t.} & x \in X \\ & x \in \{0, 1\}^{|O||L|}, \end{array} \tag{4}$$

where w_1, w_2 are nonnegative weights expressing the preferences of the decision maker such that $w_1 + w_2 = 1$.

In general, it is possible to show that every solution of problem (4) is a weak Pareto optimum for the original bi-objective problem. Moreover, if the weights w_1, w_2 are both strictly positive, then every solution of problem (4) is a Pareto optimum for (1) (see again [12]). Observe that (4) is nothing else but a Linear Assignment Problem (LAP) (see [5]), and it is well-known that for such problems both the original integer formulation and its continuous linear relaxation have the same (integer) solutions. Hence, to solve (4), one can apply any continuous linear programming technique to its linear relaxation. We omit the rather standard details for the sake of brevity.

4 The Management Software for OMNIA

The implementation of the management software devised for OMNIA includes both the strategy of maximizing the hypervolume and the more classical one of optimizing the objectives' weighted-sum. In fact, as requested by the social co-operative, the software provides three different Pareto optimal configurations (workers-jobs assignments), each one with its own peculiarities:

(a) the one that takes both the objectives into account in a "balanced" way, following the hypervolume strategy (maximization of the social well-being of both the workers and the recipients);
(b) the one that favors the quality of the service provided (maximization of the well-being of the recipients);
(c) the one that focuses on the transportation costs for the workers (maximization of the well-being of the workers).

As for case (a), the values $f_1^{\max}$, $f_2^{\max}$ have been computed by solving two linear programming problems consisting in the maximization of f_1 and of f_2, respectively.

The Pareto optimum is computed, as highlighted before, by a branch-and-bound algorithm for the MINLP (3). The solutions for cases (b) and (c) are determined by the weighted-sum method, therefore by addressing a linear programming problem: a weight of 0.99 is used for the emphasized objective, while the value 0.01 for the remaining one. We stress again (see also Sect. 3) that, according to the needs of OMNIA, the software is not intended to recover all the Pareto front, but to compute, in an a priori approach fashion, Pareto optima enjoying distinctive features.

The management software is implemented in `C++`. The `COIN-OR` suite available at www.coin-or.org is used for the solvers of the optimization problems. In particular, `CLP` and `Ipopt` are employed as linear and nonlinear programming solvers, respectively. All these packages are freely distributed under the Eclipse Public License 1.0 license. The MINLP is solved by adopting the branch-and-bound procedure from [4] and using `Ipopt` with default options to address the nonlinear relaxations.

Depending on the Pareto optimal configuration adopted by the decision maker, the final product consists in a schedule for the workers that can be directly employed to organize the provided social services as best as possible.

Currently, the software is profitably used by the OMNIA planners to organize the shifts of the workers. The planners tend to mainly use the method without preferences, since in many cases they are not able to accurately tune the trade-off between the two objectives. Nonetheless, specific cases arise not so rarely in which one of the two objectives is significantly preferred to the other. In such cases the weighted-sum method is adopted.

References

1. Auger, A., Bader, J., Brockhoff, D., Zitzler, E.: Theory of the hypervolume indicator: optimal distributions and the choice of the reference point. In: Proceedings of the 10th ACM SIGEVO Workshop on Foundations of Genetic Algorithms, pp. 87–102. ACM (2009)
2. Avriel, M., Diewert, W., Schaible, S., Zang, I.: Generalized Concavity. SIAM, Philadelphia (2010)
3. Bazaraa, M., Sherali, H., Shetty, C.: Nonlinear Programming, Theory and Algorithms, vol. 2. Wiley, Hoboken (1993)
4. Belotti, P., Kirches, C., Leyffer, S., Linderoth, J., Luedtke, J., Mahajan, A.: Mixed-integer nonlinear optimization. Acta Numer. **22**, 1–131 (2013)
5. Burkard, R., Dell'Amico, M., Martello, S.: Assignment problems, revised reprint. Siam **106** (2012)
6. Burkard, R.E.: Quadratic assignment problems. Eur. J. Oper. Res. **15**(3), 283–289 (1984)
7. Burkard, R.E.: Quadratic assignment problems. Handbook of Combinatorial Optimization, pp. 2741–2814. Springer, Berlin (2013)
8. Carraresi, P., Malucelli, F.: A new lower bound for the quadratic assignment problem. Oper. Res. **40**(1–supplement–1), S22–S27 (1992)
9. Cesarone, F., Lampariello, L., Sagratella, S.: A risk-gain dominance maximization approach to enhanced index tracking. Financ. Res. Lett. **29**, 231–238 (2018)
10. Fleischer, M.: The measure of Pareto optima applications to multi-objective metaheuristics. Evolutionary Multi-criterion Optimization, pp. 519–533. Springer, Berlin (2003)
11. Lootsma, F., Athan, T., Papalambros, P.: Controlling the search for a compromise solution in multi-objective optimization. Eng. Optim.+ A35 **25**(1), 65–81 (1995)

12. Miettinen, K.: Nonlinear Multiobjective Optimization, vol. 12. Springer Science & Business Media, Berlin (2012)
13. Zitzler, E., Thiele, L.: Multiobjective optimization using evolutionary algorithms: a comparative case study. Parallel Problem Solving from Nature-PPSN V, pp. 292–301. Springer, Berlin (1998)

Transportation

An Algorithm for the Optimal Waste Collection in Urban Areas

Edoardo Fadda, Guido Perboli and Roberto Tadei

Abstract In waste collection, one of the most important decision regards the routing and the scheduling of the service. In this context, during the Optimization for Networked Data in Environmental Urban Waste Collection (ONDE-UWC) project, an innovative optimization method for tackling those decisions has been developed in collaboration with the company Cidiu S.p.A (www.cidiu.to.it). The importance of the method is three-folds. First, it is innovative because it does not impose periodic routes. Second, it uses information coming from IoT sensors in order to build statistical models for the waste collection evolution. Third, the developed method has shown great usability and performance in the real field.

Keywords Waste management · Scheduling heuristic · Routing · Heuristic

1 Introduction

Waste collection in urban areas is a challenging problem due to urbanization and to consumption growth. The citizens of the European Union generate more than 2.3 billion tons of refuse each year. Although municipal waste represents only about 10% of total waste [5], it is of central importance for city livability. Furthermore, the efficiency of the collection services directly influences the tax amount, the emissions of pollutant, the health of citizens and visual pollution. Thus, improving the efficiency of waste management results in substantial advantages for the whole community.

In this context, this paper presents the results of the project Optimization for Networked Data in Environmental Urban Waste Collection (ONDE-UWC) (http://onde.city). This project was funded by the Regional Council of Piemonte. The importance

E. Fadda (✉) · G. Perboli · R. Tadei
DAUIN, Politecnico di Torino, Torino, Italy
e-mail: edoardo.fadda@polito.it

G. Perboli
e-mail: guido.perboli@polito.it

R. Tadei
e-mail: roberto.tadei@polito.it

M. Dell'Amico et al. (eds.), *A View of Operations Research Applications in Italy, 2018*, AIRO Springer Series 2, https://doi.org/10.1007/978-3-030-25842-9_12

of the proposed approach is twofold. First, while in the literature the waste collection problem is usually considered in the periodic settings (i.e., the dumpsters are forced to be visit periodically [9, 11]), the proposed approach removes this constraint. Second, the heuristic developed finds good solutions in a reasonable amount of time. For these reasons, the proposed approach has been adopted by the Cidiu S.p.A IT system.

The present paper is an extension of [6] including the exposition of the statistical model and the description of the IoT effects on the mathematical model. It is organized as follows: in Sect. 2, the waste collection problem is described, in Sect. 3 we revise the literature of the problem. In Sect. 4 we present the mathematical model, in Sect. 5 we describe the solution approach and in Sect. 6 we show the numerical results. Finally, in Sect. 7 we present the conclusion of the work and we depict future lines of research.

2 Problem Description

Cidiu S.p.A. is a company dealing with environmental-services. It collects municipal waste in the towns of Alpignano, Buttigliera Alta, Collegno, Druento, Grugliasco, Pianezza, Rivoli, Rosta, San Gillio, and Venaria Reale (Fig. 1). It is responsible for the collection of five types of waste: paper, solid urban refuse, plastic, metallic materials, and glass.

Cidiu S.p.A. is organized in two independent headquarters, one in Rivoli and one in Collegno. Each one of them manages the waste-collection operations in its area without interacting with the other one.

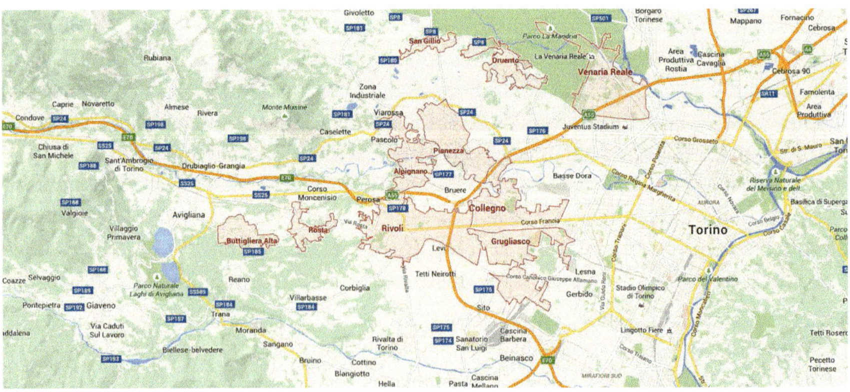

Fig. 1 The Municipalities Served by Cidiu S.p.A. Are Geographically Dispersed. Reprinted by permission, E. Fadda L. Gobbato, G. Perboli, M. Rosano, R. Tadei, "Waste Collection in Urban Areas: A Case Study", Interfaces, 48, 4, 2018. Copyright 2018, the Institute for Operations Research and the Management Sciences, 5521 Research Park Drive, Suite 200, Catonsville, Maryland 21228 USA

From Cidiu S.p.A. experience, each dumpster must be voided before it is 80 percent full.

The fleet of vehicles used to collect the waste is composed by 8 trucks (4 for each headquarter). Each vehicle must start its activity from one headquarter, travel to a predetermined set of dumpsters from which it collects waste, go to the dump, and return to the headquarter.

Each vehicle is equipped with an IoT sensor able to measure the weight of the dumpsters that it collects. This information can be exploited in several ways. First, it is possible to develop a statistical model for forecasting the daily increment of the waste quantity in each dumpster. Second, it is possible to monitor the production of waste and to react to sudden increments of production (e.g. the arrival of more people in a zone of the city).

Finally, it is possible to share the data with the population in order to improve the awareness of the waste collection activities. It is worthwhile noting that the vehicles are equipped with other IoT sensors such as the GPS tracking, sensors of the vehicle state, etc. Nevertheless, we do not consider those sensors because they do not provide information useful for the optimization problem.

Cidiu S.p.A. organizes its operations in three time shifts. The first two are done in the morning (from 6 a.m. to 12 a.m.) and in the afternoon (from 1 p.m. to 7 p.m.); they have the same cost. Instead, the third time shift is more expensive and it is done during the night (from 10 p.m. to 4 a.m.). For this reason, one of the principal concern of Cidiu S.p.A. is to reduce the usage of the third time shifts.

The decisions that we want to optimize are how many vehicles to use in each time shift, which type of waste to collect and the list of dumpsters to visit.

The computation of the routing and scheduling must be computed in less than 6 h (the duration of a time shift), in this way the Cidiu S.p.A. technical staff can check the solution and send it to the drivers.

It is important to remark that the problem description was obtained by using the GUEST methodology. This methodology allows people with different background to share their knowledge in a fast and easy way. We do not describe this methodology here because it is out of the topic (the interested reader is referred to [6]).

3 Literature Review

The literature about waste collection and operation research is very wide. To the authors knowledge one of the most recent literature review about the topic is [7]. It classifies the literature in strategic, tactical, and operational decisions by considering the time horizon of the decisions. Strategic decisions deal with long range decisions such as building new facilities or buying new vehicles. Tactical decisions deal with medium range decisions such as dumpster locations. Instead, operational decisions deal with short range decisions such as the choice of day-to-day activities. The present paper belongs to this last branch. Due to its importance several works in this field have been done. The main stream of the literature about waste collection focuses in two problems: the computation of optimal routes for vehicles (see [4, 11]) and the research for the optimal emptying frequency (see [8, 9]). In particular, the papers dealing with this second problem compute the optimal emptying frequency of each dumpster and then they use this information in order to solve a periodic capacitated arc routing problem (e.g. [3]).

As stated above, we consider the problem of computing the scheduling and routing of waste collection, without imposing a fixed voiding frequency. Thus, we allow the solutions to define the time of a most suitable next visit to empty each dumpster. This approach has received less attention because companies often have agreements with the municipalities that involve fixed frequencies and because aperiodic solution needs a model of the growth rate of the waste in each dumpster.

Although the aperiodic approach is less explored in the literature, another study ([10]) not considering fixed routes is available. It presents a heuristic for applying the inventory-routing problem to waste collection, considering data from sensor-equipped containers.

Our study and that of [10] are different for several reasons. First, [10] considers a mathematical model addressing only vehicle routing decisions. Instead, our approach considers also the scheduling of the operations. Second, [10] uses sensor-equipped containers to forecast the level of fullness of each dumpster. This is, from the application point of view, a risky choice because the company must do the maintenance of all the dumpsters. Instead, having the sensors on the vehicles is more reliable and less subject to fault.

4 Mathematical Formulation and Solution Method

Let $\mathscr{I}$ be the set of available vehicles, $\mathscr{S}$ the set of garbage types, $\mathscr{T}$ the set of time shifts and $\mathscr{J}$ the set of dumpsters. The cardinalities of these sets are I, S, J, and T, respectively. Without loss of generality we consider that the set $\mathscr{J}$ also includes the depot ($j = 1$) and the dumps ($j = J - S, \ldots, J$ one for each type of waste). We use the following parameters:

- c_{it} is the cost of using vehicle $\forall\ i \in \mathscr{I}$ during shift $t \in \mathscr{T}$. This cost is the same for the first two time shifts and greater for the third shifts.
- $C_{\max}$ is the maximum duration of the time shift.
- $d_{j_1 j_2}$ is the time that the vehicle requires to go from dumpster $j_1 \in \mathscr{J}$ to dumpster $j_2 \in \mathscr{J}$.
- $\hat{C}$ is the maximum volume that a vehicle can transport.
- l_s is the difference between a standard quantity that a vehicle can transport and the real quantity if it is used to collect waste type $s \in \mathscr{S}$. This difference results from the use of a press (i.e., an engine that can compress the waste).
- Θ_{jt} is the increment of volume of waste in dumpster $j \in \mathscr{J}$ during time $t \in \mathscr{T} - 0$; Θ_{jt} is the quantity of waste in dumpster $j \in \mathscr{J}$ at the beginning of the first time shift.

- a is the time that the driver needs to collect the waste from a dumpster. Because of the technology installed on the vehicles, the collection time is the same for all dumpsters.
- α is the percentage of the dumpster volume that cannot be exceeded.
- λ is a parameter that weighs the importance of the routing with respect to the operational costs. It is expressed in €$/h$ and it considers the fuel cost and the manpower cost of the trip per hour.
- V_j is the capacity of dumpster j.

It is worthwhile noting that the parameters Θ_{jt} $\forall\, t$ are computed by using the statistical models of the growth rate of the waste in each dumpster j. These models have been fitted by using the data recorded from the IoT sensors installed on the vehicles.

The following variables are used:

- w_{it} binary variables assuming value one if vehicle $i \in \mathscr{I}$ is used during shift $t \in \mathscr{T}$
- z_{ist}, binary variables assuming value one if vehicle $i \in \mathscr{I}$ collects the garbage of type $s \in \mathscr{S}$ during shift $t \in \mathscr{T}$,
- y_{ijt}, binary variables assuming value one if vehicle $i \in \mathscr{I}$ collects the garbage of dumpster $j \in \mathscr{J}$ during shift $t \in \mathscr{T}$
- $r^{it}_{j_1 j_2}$, binary variable assuming value one if vehicle $i \in \mathscr{I}$ during time shift $t \in \mathscr{T}$ goes from dumpster $j_1 \in \mathscr{J}$ to dumpster $j_2 \in \mathscr{J}$.
- x_{ijt}, continuous variables describing the volume of garbage collected by vehicle $i \in \mathscr{I}$ from dumpster $j \in \mathscr{J}$ during shift $t \in \mathscr{T}$,
- V_{jt}, continuous variables describing the volume of waste present in dumpster $j \in \mathscr{J}$ at the end of time shift $t \in \mathscr{T}$.

The mathematical problem describing Cidiu's activities is:

$$\min \sum_{i=1}^{I} \sum_{t=1}^{T} c_{it} w_{it} + \lambda \sum_{i=1}^{I} \sum_{t=1}^{T} \sum_{j_1=1}^{J} \sum_{j_2=1}^{J} d_{j_1 j_2} r^{it}_{j_1 j_2} \tag{1}$$

subject to

$$w_{it} \geq y_{ijt} \quad \forall\ i, j, t \tag{2}$$

$$M y_{ijt} \geq x_{ijt} \quad \forall\, i, j, t \tag{3}$$

$$\sum_{i=1}^{I} y_{ijt} \leq 1 \quad \forall\, j, t \tag{4}$$

$$x_{ijt} \geq V_{jt} - M(1 - y_{ijt}) \quad \forall\, i, j, t \tag{5}$$

$$\sum_{j=1}^{J} x_{ijt} \leq \hat{C} + \sum_{s=1}^{S} l_s z_{ist} \quad \forall\ i, t \tag{6}$$

$$Mz_{ist} \geq x_{ijt} \ \ \forall\, i, s, j \in \mathscr{J}^s, t \tag{7}$$

$$\sum_{s=1}^{S} z_{ist} \leq 1 \ \ \forall\, i, t \tag{8}$$

$$V_{j0} = \Theta_{j0} \ \ \forall\, j \tag{9}$$

$$z_{ist} \leq y_{i(J-s)t} \ \ \forall\, i, t \tag{10}$$

$$V_{jt} = V_{jt-1} + \Theta_{jt} - \sum_{i=1}^{I} x_{ijt} \ \ \ \forall\, j, t \neq 0 \tag{11}$$

$$V_{jt} \leq \alpha V_j \ \ \ \forall j, t \tag{12}$$

$$w_{it} \geq w_{i+1,t} \ \ \forall\, t, i = 1 : I - 1 \tag{13}$$

$$\sum_{j_1=1}^{J} r_{jj_1}^{it} = y_{ijt} \ \ \forall\, t, i, j \tag{14}$$

$$\sum_{j_1=1}^{J} r_{jj_1}^{it} = \sum_{j_1=1}^{J} r_{j_1 j}^{it} \ \ \forall\, t, i, j \tag{15}$$

$$\sum_{j=1}^{J} r_{1j}^{it} = w_{it} \ \ \forall\, t, i \tag{16}$$

$$\sum_{j=1}^{J} r_{j1}^{it} = w_{it} \ \ \forall\, t, i \tag{17}$$

$$\sum_{j_1=1}^{J} \sum_{j_2=1}^{J} d_{j_1 j_2} r_{j_1 j_2}^{it} + a \sum_{j=1}^{J} y_{ijt} \leq C_{\max} \ \ \forall\, i, t \tag{18}$$

$$\sum_{j_1, j_2 \in S, j_1 \neq j_2} r_{j_1 j_2}^{it} \leq |S| - 1 \ \ \forall S \subset \mathscr{J}, S \neq \emptyset \tag{19}$$

$$x_{ijt} \in \mathbb{R}^+ \ \ \ \forall i, j, t$$

$$z_{ist} \in \{0, 1\} \ \ \ \forall i, s, t$$

$$y_{ijt} \in \{0, 1\} \ \ \ \forall i, p, t$$

$$r^{it}_{j_1 j_2} \in \{0, 1\} \quad \forall i, j_1, j_2, t$$

$$V_{jt} \in \mathbb{R}^+ \quad \forall j, t$$

The objective function is the weighted sum of the costs derived from the usage of a vehicle during a time shift and the routing cost. Constraints (2) and (3) are logic constraints enforcing that it is not possible to use a vehicle during a time shift, without activating the corresponding time shift. Constraints (4) enforce that, in each time shift, no more than one vehicle can void a dumpster. Constraints (5) enforce that the vehicles must collect all the garbage from the dumpsters it visits. Constraint (6) enforces the capacity limit of the vehicles; the right hand side considers that for some types of waste the vehicle can press the waste, thus diminishing the volume. Constraints (7) and (8) enforce that vehicles cannot collect waste from dumpsters of different type. Constraints (10) enforce that if a vehicle is used for voiding waste type $s \in \mathscr{S}$, it must go to the assigned dump. Constraints (9), (11), and (12) enforce the model of the evolution of the waste in each dumpster and bound this quantity. Constraints (13) enforce an order in the vehicle choice. These constraints are used for breaking the symmetry of the problem by reducing the feasible set. Finally, constraints (14)–(19) describe the routing problem and set a maximum time for the activity.

It is worth noting that because of constraints (9) and (11) the value of V_{jt} can be modified only by using the variables x_{ijt}. Furthermore, if variables variables x_{ijt} are null, then constraint (12) is violated because the contributions $\Theta_{jt} > 0 \forall j, t$.

4.1 *The Statistical Model*

For each dumpster, the company records the waste quantity collected. We call $\hat{\theta}_{jn}$ the observed quantity of waste collected from dumpster j during the collection n, θ^p_{jn} the corresponding random variable, and t_n the time (expressed in number of time shift) of the n-th collection. Given a waste producer p belonging to the set of waste producers $\mathscr{P}$, we call the quantity of waste produced by p in dumpster j and collected during the collection n as θ^p_{jn}. It holds that

$$\theta_{jn} = \sum_{p \in \mathscr{P}} \theta^p_{jn} \tag{20}$$

We assume that the quantity of waste observed in each dumpster is produced uniformly in time. Furthermore, we suppose that the distributions of the $\hat{\theta}^p_{jn}$ have finite mean and variance and that the random variables $\hat{\theta}^p_{jn}$ are independent. Furthermore, we assume that the Lyapunovs condition holds, i.e., there exists $\delta > 0$ such that

$$\lim_{n \to \infty} \frac{1}{s_n^{2+\delta}} \sum_{i=1}^{n} \mathbb{E}\left[|\theta^p_{jn} - \mu_{jn}|^{2+\delta} \right] = 0, \tag{21}$$

where μ_{jn} and s_n are respectively the average and the square root of the sum of the variances of the θ^p_{jn}.

The Lyapunovs condition guarantees that the moments of the distributions do not increase too much as it is reasonable to assume. By applying the Lyapunov central limit theorem [2], we have that the θ_{jn} are normal distributed random variables. Thus, we use as estimator for the production rate the average daily fulfillment rate:

$$\Theta_{jt} = \frac{1}{N}\sum_{n=1}^{N} \frac{\hat{\theta}_{jn}}{(t_n - t_{n-1})} \quad \forall j, t.$$

5 The Solution Algorithm

The mathematical model described in (1)–(19) is a mixed-integer linear problem. Thus, the main method to solve the problem exactly is the branch and bound algorithm. Since the number of variables of the model is $\mathscr{O}(2^{IJ^2T})$ (where I is the number of vehicles, J is the number of dumpsters, and T is the number of time steps) this is also the worst case complexity of the model. It is easy to see that even for small-size problems the number of variables is big. In order to overcome this problem we implement a heuristic method. Based on numerous experiments, we decided to use a three-phased heuristic. We describe the first phase is Sect. 5.1 and the two remaining phases in Sect. 5.2.

5.1 Clusterization

Phase 1 solves the exact problem by considering aggregation of dumpsters (called clusters), instead of single dumpster. The policy chosen for aggregation is geographical proximity i.e. we group together all the dumpsters located in the same city. This choice greatly outperforms other kinds of policy such as grouping together the dumpsters by level of productivity, last visit date, etc. We cannot present the comparison between the different methods in this paper for length constraints. It is worthwhile noting that, in order to solve the exact problem with clusters instead of dumpsters, we have to modify model (1)–(19). In particular, instead of the set $\mathscr{J}$, we consider the set $\mathscr{C}$, the set of clusters (with cardinality C). We remove constraints (4) and (5), because we allow more than one vehicle to collect waste from the same cluster. This choice increases the size of the feasible solution space and ensures the feasibility of the problem. In fact, since we consider a low number of clusters and a vehicle has enough time to visit at least one cluster, the worst feasible solution is that each vehicle collects exactly one cluster.

Moreover, in the new model the parameter $d_{c_1c_2}$ becomes the distance between cluster c_1 and cluster c_2, Θ_{ct} becomes the maximum growth rate of the dumpsters in the cluster, V_c the capacity of the cluster and $\hat{a}_C$ becomes the time spent to empty a cluster. We define them as follows:

1. $d_{c_1c_2} = \min_{j_1 \in c_1 j_2 \in c_2} d_{j_1 j_2}$;
2. $\Theta_{ct} = \max_{j \in c} \Theta_{jt}$;
3. $V_c = \max_{j \in c} V_j$. V_c is the maximum capacity of the dumpsters in the cluster;
4. $\hat{a}_C = T_v(C) + T_r(C)$ where $T_v(C)$ is the sum of the emptying time of all dumpsters and $T_r(C)$ is the time of a path among all dumpsters in the cluster.

Remark 1 It is worthwhile noting that the parameter $\hat{a}_C$ used in the cluster model is the sum of the time for visiting all dumpsters in the clusters plus their emptying time. In other words, it represents the worst possible case i.e. the one in which all dumpsters in the clusters must be voided. We choose this very conservative value because, in the cluster model, $d_{c_1c_2}$ is a lower approximation for the travelling time and by that choice we ensure that for each travelled edge (c_1, c_2) the sum $\hat{a}_{C_1} + d_{c_1c_2} + \hat{a}_{c_2}$ is an upper bound of the total time spent for the collection.

On average, the trip time computed in the cluster model is 1.5 times the one computed in the real model. Nevertheless, since the critical resource of the problem is the volume of the vehicle, this approach does not remove any feasible solution from the solution space.

5.2 Building a Feasible Solution and Postoptimization

By using the solution computed in the Phase 1, Phase 2 builds a feasible solution for problem (1)–(19). Its output is, for each vehicle and for each time shift, a list of dumpsters that the vehicle must void. The procedure is shown in the following Algorithm 1.

Algorithm 1 Post Optimization

Data: Solution of the model (1)–(19) for the cluster network
Results: Solution of the model (1)–(19) for the real network
for each $w_{it} = 1$, where $i \in \mathcal{I}$, $t \in \mathcal{T}$ **do**
  create list *list0*
  insert in *list0* all dumpsters for each cluster, such that $y_{ict} = 1$ where $i \in \mathcal{I}$, $c \in \mathcal{C}$, $t \in \mathcal{T}$
  remove from *list0* all dumpsters that have been voided in the previous iterations
  sort *list0* in decreasing order of quantity of waste
  create *list1*
  while Constraints (6) or (18) computed on *list1* hold with equality and *list0* $\neq \emptyset$ **do**
    add, the first element of *list0* into *list1*
    remove the first element of *list0*
  end while
  if *list0* $= \emptyset$ **then**
    add all the dumpsters that are such that if voided the time of the path as well as the capacity of the vehicle is not exceed
  end if
end for

From the solution provided by Algorithm 1, Phase 3 optimizes the routing. In particular, for each list of dumpsters that a vehicle has to void during a time shift, it computes the solution of a traveling salesman problem (TSP) by using the Chained-Lin-Kernighan heuristic for asymmetric networks, as [1] discusses.

Because the Phase 1 and Phase 2 run quickly, we run them several times by randomly changing the order in which the dumpsters in *list0* are ordered.

Remark 2 It is worthwhile noting that the variables z_{ist} are set by the association of a vehicle to a type of waste and they are derived from the exact solution of the clusterized model.

6 Computational Experiments

In order to evaluate the solutions provided by the heuristic, we considered six key performance indicators (KPIs):

- nTS: average number of third shifts used during a week of activity.
- nV: average number of vehicles daily used. This KPI is the ratio between the number of vehicles used during one month of activity and the number of days.
- WV%: average percentage of waste volume at the moment of collection. During the collection operations, the vehicle records the volume occupied by the waste. The indicator is computed by averaging the percentage of occupation over all the collections and all the dumpsters collected;
- TRT: total routing time.
- FV%: average fulfillment percentage of the vehicles. This KPI is the average percentage of fulfillment over each time shift and over each vehicle.
- nVD: average number of dumpsters visited during each time shift in which a vehicle is used.

In Table 1 we show the KPIs values for a real month of activity for Cidiu S.p.A. and a simulated month of activity for the proposed method (within brackets the standard deviation of each value is shown). The simulated month is the same as the real one.

Table 1 For each KPI, we compare the solution before and after using our proposed algorithm

KPIs	Cidiu S.p.A. solution	Simulated solution
nTS	1.44 (0.5)	0 (0)
nV	3 (0)	2 (0)
WV%	0.28 (0.10)	0.70 (0.05)
TRT	4.35 (0.5)	5.24 (0.3)
FV%	54% (10%)	87% (5%)
nVD	62.3 (12)	68.5 (12)

The results presented in Table 1 show that the proposed methodology outperforms the actual policy with respect to all the KPIs. The most important result is that the proposed method do not use the third shift. Furthermore, it uses 33% less vehicles. It is worthwhile noting that the increase in the total routing time is due to the usage of fewer vehicles that have to travel more. For this reason, the growth of the total routing time is not due to inefficiency but it denotes a better usage of the vehicles.

In order to test the effects on the real field of the algorithm, Cidiu S.p.A. uses the methodology for small periods of time obtaining similar results to the ones in the simulation. Currently, due to the good results, Cidiu S.p.A. is integrating the proposed methodology into its IT system.

It is important to note that the problem depends by the parameter λ that converts the time of the routing in cost for the company. Unluckily, it is difficult to estimate because it is generated by fuel consumption, the cost of the drivers, the depreciation of the vehicle, etc. Nevertheless, variations in the value of this parameter change slightly the solution. Another parameter that does not influence the solution is the capacity of the vehicle $\hat{C}$. This is due to the fact that the most critical resource is the duration of the time shift and not the volume of the vehicle.

Instead, the model is sensitive to changes in the Θ_{jt}. The greater those quantities the greater the total cost. It is worthwhile noting that for Θ_{jt} sufficiently high the model becomes unfeasible. By this mathematical observation, it is possible to deduce the importance of the data recorded by the IoT sensors measuring the weight of the refuse. An accurate estimation of the waste collection production leads the model to be really effective. Indeed, this is an added value of the proposed methodology that, to the authors knowledge, has never been considered in the literature.

Another parameter that deeply influences the model is the duration of the time shift $C_{\max}$. The greater this value, the lower the objective function.

In order to estimate the optimality gap of the solutions obtained by the proposed heuristic, we compared them with the optimal solution found using CPLEX 12.6. Since the solver is not able to solve real instances because it runs out of memory, we consider instances with 10, 20, 30, 40, and 50 dumpsters, six time periods, and two types of waste.

In those instances, the dumpsters are spatially generated from a combination of multinomial distributions, in order to be located in scattered ellipses (as in the real settings). The initial waste quantity of each dumpster and its increment are simulated from the historical data. The travel times are computed by using the real distances and the costs of each time shift are the same as the one of Cidiu S.p.A.

Table 2 shows the results of this comparison, we report the differences between the costs of time shifts and the costs of the routing of the two solutions. In each cell the table shows the average value over 20 runs and standard deviation (in brackets).

It is worthwhile noting that the proposed method finds the optimal number of time shifts in all the generated instances. The differences between the optimal solution and the one found by the heuristic are in the routing.

Furthermore, the average running time of the proposed method in the real setting is 4 h and 23 min with a standard deviation of 20 min (these values are computed by using 100 simulations).

Table 2 The table shows a comparison of Cidiu S.p.A.'s optimal solutions and the solutions generated using our proposed approach

Number of dumpsters	nTs [%]	rC [%]	Optimal time [s]	Heuristic time [s]
10	0 (0)	0 (0)	43.43 (1.64)	2.76 (0.65)
20	0 (0)	0 (0)	150.43 (10.89)	3.73 (0.56)
30	0 (0)	1.75 (1.38)	443.64 (15.92)	8.56 (0.84)
40	0 (0)	2.69 (1.43)	890.67 (30.53)	13.36 (0.74)
50	0 (0)	3.32 (2.34)	1842.86 (45.65)	26.27 (2.45)

Thus, the proposed heuristic can be run once every time shift, allowing the management to adjust the plan and to consider missed operations (e.g., the vehicle cannot collect the waste because a car is parked near the dumpster).

7 Conclusions

In conclusion, we claim that the proposed method has proven to be effective in the real field: the system presented is currently being integrated into the IT system of Cidiu S.p.A. and, according to the company forecasts, the system will be fully operational by the end of 2018. By doing so, we have proved that the aperiodic approach is worth to be studied and that heuristics dealing with this problem can be useful in real world.

References

1. Applegate, D., Cook, W., Rohe, A.: Chained lin-kernighan for large traveling salesman problems. INFORMS J. Comput. **15**(1), 82–92 (2003)
2. Billingsley, P.: Probability and Measure. Wiley, New York, NY, USA (1995)
3. Chu, F., Labadi, N., Prins, C.: A scatter search for the periodic capacitated arc routing problem. Eur. J. Oper. Res. **169**(2), 586–605 (2006)
4. Das, S., Bhattacharyya, B.: Optimization of municipal solid waste collection and transportation. Waste Manag. **43**, 9–18 (2015)
5. European Environment Agency: http://www.eea.europa.eu/soer-2015/countries-comparison/waste (2015). Accessed 27 July 2016
6. Fadda, E., Gobbato, L., Perboli, G., Rosano, M., Tadei, R.: Waste collection in urban areas: a case study. Interfaces **48**(1), 103–120 (2018)
7. Ghiani, G., Laganà, D., Manni, E., Musmanno, R., Vigo, D.: Operations research in solid waste management: a survey of strategic and tactical issues. Comput. Oper. Res. **44**, 22–32 (2014)
8. Hemmelmayr, V., Doerner, K., Hartl, R., Vigo, D.: Models and algorithms for the integrated planning of bin allocation and vehicle routing in solid waste management. Informs Transp. Sci. **48**(1), 103–120 (2013)
9. Mansini, R., Speranza, M.: A linear programming model for the separate refuse collection service. Comput. Oper. Res. **25**(8), 659–673 (1998)

10. Mes, M., Schutten, M., Rivera, A.: Inventory routing for dynamic waste collection. Waste Manag. **34**(9), 1564–1576 (2014)
11. Sahoo, S., Kim, S., Kim, B., Kraas, B., Popov, A.: Routing optimization for waste management. Interfaces **35**(1), 24–36 (2005)

A Decision Support System for Earthwork Activities in Construction Logistics

Mauro Dell'Amico, Guenther Fuellerer, Gerhard Hoefinger and Stefano Novellani

Abstract Making decisions in a complex system such as the construction of a highway is a hard task that involves a combinatorial set of possibilities, concerning thousands of interrelated activities and resources over several years. In this paper we describe a decision support system (DSS) developed to assist project managers in decision making for the construction of the Autostrada Pedemontana Lombarda highway, in Italy. The considered problem evaluates the earthwork activities in detail and defines the minimum cost earthwork plan satisfying all constraints. The proposed DSS involves the use of linear programming to solve the earthwork problem in a two-phase approach: in the first phase, an aggregate model determines the feasibility of the overall project, whereas in the second phase, disaggregate models determine the actual flows of each material. The DSS gathers the needed information directly from the master plan commonly used by the company and provides as output a set of visual solutions. The solution are yielded in short times and can be run many times with different data sets supporting a fast evaluation of different decisions. The provided solutions are also optimized and could substitute the previous manual results. The DSS has been proved to be very effective for assisting the project managers of the above highway construction and is currently in use in other projects.

Keywords Decision support system · Earthwork · Construction logistics · Mathematical modeling

M. Dell'Amico (✉) · S. Novellani
DISMI, Università di Modena e Reggio Emilia, Reggio Emilia, Italy
e-mail: mauro.dellamico@unimore.it

S. Novellani
e-mail: stefano.novellani@unimore.it

G. Fuellerer · G. Hoefinger
STRAsco Team, Zentrale Technik, Strabag AG, Vienna, Austria
e-mail: guenther.fuellerer@strabag.com

G. Hoefinger
e-mail: gerhard.hoefinger@strabag.com

M. Dell'Amico et al. (eds.), *A View of Operations Research Applications in Italy, 2018*, AIRO Springer Series 2, https://doi.org/10.1007/978-3-030-25842-9_13

1 Introduction

A construction project is a complex system that may involve hundreds or thousands of interrelated activities, resources and operators, and may have a duration of several years. To manage and coordinate the execution of the activities, project management software applications have been typically used. These tools assist the project responsible to plan the sequence of the activities, to monitor the progress of the execution, and to define reactions to deviations from the original plan. A project management software is usually independent from software tools used for pure design and for structural analysis. Other software applications are possibly used to design the hydraulic system, the electrical system, the heating system etc. In the last fifteen years it became popular the concept of Building Information Modeling (BIM) which aims to represent in a sole system all the information needed for the construction design and management. BIM tools appeared in the noughties and have been gradually adopted by several large and medium size construction companies. More recently, some information for project management has been added to the BIM software. BIMs represent a large innovation that allows to have all information on the project in the same system, making easier to make decisions. However, BIMs do not incorporate optimization algorithms able to help the manager to find the best planning when alternative choices and solutions exist. In this case the user adopts a scenario based strategy which consists in listing a set of different alternatives, evaluating each of them with the planning tool and then selecting according to the measured performances. This method can give reasonable solutions when there are few alternatives, but becomes impractical when several choices exist, due to the combinatorial explosion of the possibilities. The second aspect that cannot be effectively handled is the management of the operational phase, where a very large set of small activities have to be considered.

In this paper we present a Decision Support System (DSS) designed for the management of earth movements and operations planning in a large highway construction project. Consider that *earthwork* operations account for about 25% of the construction costs of an highway (see [10]). The DSS has been successfully used during the construction of the Autostrada Pedemontana Lombarda (Italy) in the years 2012–2016 and is currently used in several other projects by Strabag AG. In Sect. 2 we give a general introduction on the real life problem that inspired the work, while in Sect. 3 we shortly resume the relevant literature. Section 4 presents in more detail the elements and constraints of the problem and introduces the approach we designed. Section 5 explains the mathematical models and solution methods we implemented. Section 6 describes the complete decision support system, its use and advantages in the real life case. Section 7 summarizes our work and its effectiveness for the construction company.

2 The Construction Project

Autostrada Pedemontana Lombarda (www.pedemontana.com) is a large highway construction project part of the activities planned for the *World Exposition Milano 2015* (Expo Milano 2015). It runs North of Milan (see Fig. 1) and includes a large system of highways, access ramps, and local streets.

It involves 87 km of highway (including bypasses) and 70 km of links to ease local vehicular load. About 35% of the roads run through natural or artificial galleries. Figure 2 summarizes the characteristics of the highway. The planned cost of the project is of about 4.1 billions € and its construction time five years.

The aim of the project is to redistribute the traffic in the northern area of Milan and link the three main airports of the region. The project has a high economically and socially impact, since the highway runs through a territory with more than 220,000 companies (40% of the regional total) and four million inhabitants. The impact of the traffic over the residential areas, due to the new highway, is minimized using trench and artificial or natural galleries for 75% of the highway path. This induces a very complex earth movement problem where are excavated and moved roughly 30 millions m^3 of earth. The construction has been entrusted to a consortium leaded

Fig. 1 *Autostrada Pedemontana Lombarda*: an highway bypass north of Milan (Italy)

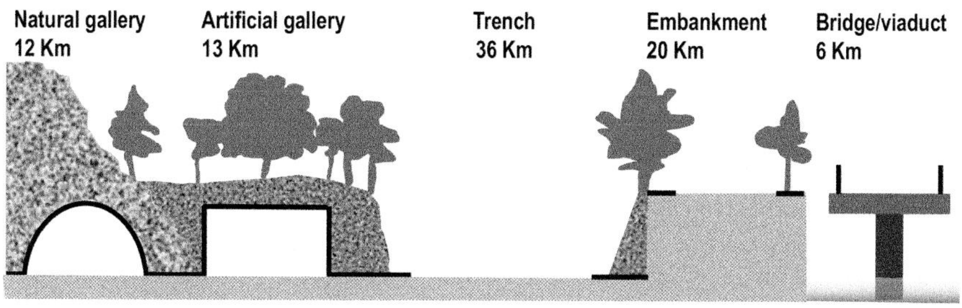

Fig. 2 The main characteristics of the highway

by the Austrian company Strabag AG, which holds 60% stake of the project, within other Italian constructors. Strabag AG Group, with headquarter in Vienna (Austria), has know-how and experience built up over more than 170 years of business activity. With approximately 73,000 employees, the group generated in the financial year 2017 a construction output volume of about € 15.0 billion. It operates in the core markets of Austria and Germany, in many countries of Eastern and South-East Europe, in selected markets in Western Europe and in individual cases on other continents. The company offers a wide range of services that span all areas of the Construction Industry: building construction and civil engineering, infrastructure construction and tunneling, and facility management.

Zentrale Technik is an Engineer-center of excellence that develops the technical know-how of the Strabag AG Group and spreads it within the company. Composed of approximately 700 employees at 18 locations, it represents a technical support for all divisions in decisions regarding acquisition and construction. The *Supply Chain Optimization* (STRAsco) team of Strabag AG, composed of experts in operations research, network optimization, simulation, and logistics, treats topics along the whole supply chain and delivers a well founded basis to decision makers in the field of construction logistics. STRAsco operates in the Zentrale Technik of Vienna since January 2013. This work is the result of the collaboration between the STRAsco group and the Operation Research group of University of Modena and Reggio Emilia.

3 Literature Review

The basic earthwork activities are described by the classic *cut and fill problem* that can be easily modeled and solved as a minimum cost flow problem (see, e.g., [1]). In [8, 9] it is investigated the use of non-linear cost functions to model purchase and excavation in borrow pits. Uncertainties in costs and facility capacities have been considered in [11] assuming a fuzzy logic.

A knowledge-based system integrated with simulation is proposed in [2] to automate the earthwork planning process. Another tool to automate the planning is proposed in [17] where several factors are integrated and genetic algorithms are used to optimize costs, productivity, safety, and environmental impacts.

Multi-objective approaches using simulation and genetic algorithms have been proposed in [12] (to improve productivity, efficiency, and safety, at the same time) in [18] (to model equipment configurations selection based on multi-attribute utility functions), and in [15] (to determine the best equipment allocation).

In [13] several tools like genetic algorithms, LP models, and GIS have been used to solve the problem while optimizing the utilization of construction equipments involved in earthmoving operations.

A mixed integer linear program (MILP) is introduced in [10] to model the fact that some earth movement can be done only after certain road sections are built. In [16] another MILP is used to optimize the excavation costs including setup costs for activating quarries and dump sites. In [14] the cut and fill problem with inclusion of equipment's sequence constraints is formulated as a mixed binary linear problem and solved by branch-and-bound and particle swarm optimization. An LP model for equipment selection in earthwork operations with a Java interface is described in [3].

A block partitioning and modelling approach is presented in [5] to allocate and sequence cut and fill operations for minimizing hauling costs. In [6] a similar approach is used to optimize fuel consumption and haulage distance and gradient. Constructive heuristics and metaheuristics such as simulated annealing and evolutionary algorithms are used to solve the problem.

A complete description of the DSS presented in this paper can be found in [4, 7].

4 Problem Details and Solution Method

The *master plan* of the highway construction is defined by the Strabag production department, using Microsoft Project. All the activities are listed and described, the precedences are drawn and a schedule is defined giving an overall Gantt chart and a *time window* for each activity. The plan is produced at the beginning of the project and continuously updated by the production department, because the original plan and time windows can be infeasible. The plan contains also information on the resources needed to perform each activity, the costs, and the capacities. The production department discretizes the construction area into segments of different length. For the aim of our problem each segment represents a *material location*. The time horizon is also discretized into *periods* and includes hundreds of intermediate milestones given by the contractor. Our problem takes as input the master plan and has to decide how to implement it, in detail. In particular we have: (i) to plan each excavation and filling activity while respecting time window constraints; (ii) to decide on the hauling of the many materials required, by using different types of vehicles running on a highly constrained network; (iii) to take decisions on buying, dumping, recycling, and storing materials. The detailed plan must minimize the overall costs.

In the basic earthwork problem, known as the cut and fill problem (see Sect. 3) one deals with the material excavation and immediate transport to a filling site. In a real construction problem the earth movements are much more complex. We may need to transport the excavated material to a temporary depot and, at a second time, to transport it to a filling site. We may transport the excavated material to a recycle site, and later transport a percentage of the material to a filling site, and the remaining to a dump site. In a filling site, if necessary or convenient we may decide to use material from *quarries*, instead of excavated material.

Due to the large quantities each "movement" is implemented through several vehicle trips. Moreover, the trips associated to a same movement are not necessarily identical, having to comply with environmental constraints which impose a limit on the number of trucks that drive on a certain roads in each period.

The excavated material must be sorted on-site to select the specific materials that can be used for construction purposes. The sorting basically identifies four kinds of material: High quality, Medium quality, Waste, and Topsoil (the outermost layer of

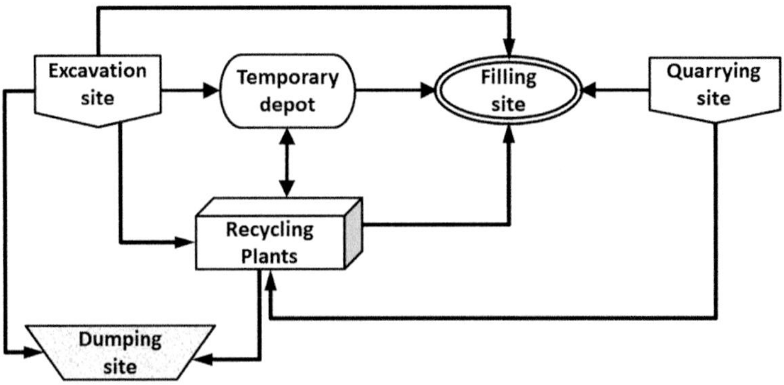

Fig. 3 Material flows in the highway construction project

soil). The percentages of the materials, for each site, are known due to preliminary geological tests. The High and Medium quality materials can be used directly for filling, but the High quality one can be also recycled to obtain a new material that can be used for construction purposes (recycling is done in specific plants where the material is mixed with *bitumen* and *cement* to obtain *concrete* and *asphalt*). Waste material must be discarded in specific *dump sites*, while Topsoil is temporarily stored and reused directly to recover the soil at the end of the construction.

Quarries, dump sites, temporary depots, and recycling plants have capacity restrictions. To move the materials, different types of capacitated vehicles are used.

Figure 3 gives a schematic representation of the main material flows of the problem.

5 A Mathematical Programming Approach

We have modelled the problem starting from a network flow, and then adding several constraints to represent all the real life needs. One part of the network represents the construction site and another part represents the external roads needed to reach dump sites, quarries, and recycle plants. The vertices of the network correspond to material locations, plants, dump sites, temporary depots, quarries, and junctions of the road network. The vertices are connected by capacitated arcs that represent the private roads, inside the construction site, and the public external roads. The external roads undergoes regulations that provide restrictions for the flows on that network.

The master plan is designed focusing on the technological constraints of the construction process, but due to the very large number of activities, of capacity constraints, of resources, and to the long duration, it is impossible to take care of all constraints at the same time, when drawing a manual solution. Indeed, we solved instances with about 1100 vertices and 2300 arcs. (In practice the company uses the master plan as a guide for the execution of the activities, then operational weekly or daily plans are made to solve the resource conflicts, capacity shortages etc. The overall result may diverge significantly from the plan, mainly in term of costs).

In a first phase the scope of our DSS is to check the feasibility of the master plan or to define a master plan update leading to a feasible solution, by means of an iterative process with the production department. Indeed, several constraints, such as the time windows, can lead the problem to infeasibility. This is the *strategic* level of the DSS. When feasibility is reached, we start a second phase aimed to minimize the overall costs (traveling, buying, dumping, and storing). This is the *operational* level of the DSS.

Both phases are based on LP models: In the first phase an *aggregate formulation* is used to check the feasibility of the project with the given timing and resources limits, while minimizing the costs. In the second phase a *disaggregate formulation* takes the solution of the aggregate formulation as input and determines the specific paths for each material and for each period of the time horizon, always minimizing the costs.

Aggregate formulation

Let us give a brief description of the aggregate formulation. The vertex set is $V = (\mathscr{P} \cup \mathscr{D} \cup \mathscr{Q} \cup \mathscr{L} \cup \mathscr{E})$ where $\mathscr{P}$ represents the temporary depots, $\mathscr{D}$ the dump sites, $\mathscr{Q}$ the quarries, $\mathscr{L}$ the material locations, and $\mathscr{E}$ the set of junction nodes to the external network. Let $\mathscr{A}$, $\mathscr{M}$, and $\mathscr{T}$ be the sets of arcs, materials, and time periods, respectively.

We use two sets of non-negative variables: x_{ij}^{mt} represents the flow of material $m \in \mathscr{M}$ in period $t \in \mathscr{T}$ on arc $(i, j) \in \mathscr{A}$; f_k^{mt} represents the quantity of material m: (a) stored in a temporary depot $k \in \mathscr{P}$; or (b) carried to a dump site $k \in \mathscr{D}$; or (c) acquired from a quarry $k \in \mathscr{Q}$, at period t.

The objective function is

$$\min \quad z = \sum_{(i,j)\in\mathscr{A}} \sum_{m\in\mathscr{M}} \sum_{t\in\mathscr{T}} c_{ij} x_{ij}^{mt} + \sum_{k\in\mathscr{P}\cup\mathscr{D}\cup\mathscr{Q}} \sum_{m\in\mathscr{M}} \sum_{t\in\mathscr{T}} c_k^m f_k^{mt}. \tag{1}$$

where c_{ij} is an estimation of transportation cost on arc (i, j) (note that in the aggregate model we do not decide on the specific type of truck that is used for each movement, so we do not know the exact transport cost), and c_k^m denote the cost for storing, dumping, or buying a unit of material m at location k.

Every arc $(i, j) \in \mathscr{A}$ has a maximum capacity U_{ij}^t for each period t, so:

$$\textstyle\sum_{m\in\mathscr{M}} x_{ij}^{mt} \le U_{ij}^t \;\; (i, j) \in \mathscr{A}, t \in \mathscr{T}. \tag{2}$$

Flow conservation at each vertex must be imposed:

$$\textstyle\sum_{(i,j)\in\mathscr{A}} x_{ij}^{mt} = \sum_{(j,i)\in\mathscr{A}} x_{ji}^{mt} \;\; j \in \mathscr{L} \cup \mathscr{E}, m \in \mathscr{M}, t \in \mathscr{T}. \tag{3}$$

An overall transportation capacity is added to help finding feasible solutions:

$$\sum_{(i,j)\in\mathscr{A}} \sum_{m\in\mathscr{M}(w)} \ell_{ij} x_{ij}^{mt} \le N(w)\cdot V(w)\cdot K(w)\ t\in\mathscr{T}, w\in\mathscr{W}, \tag{4}$$

where ℓ_{ij} is the length of arc (i, j) and W is the set of vehicle types. Each type $w \in W$ consist of $N(w)$ identical vehicles, each of which may transport a set $\mathscr{M}(w) \subseteq \mathscr{M}$ of materials, has a maximum transportation capacity $V(w)$, and a maximum number $K(w)$ of km that can be traveled.

Let $\mathscr{F}$ denote the set of *filling tasks* as defined in the master plan. Each task $i \in \mathscr{F}$ refers to material location $h(i) \in \mathscr{L}$, has a request quantity d_i, and must be performed within the time window $[t_s(i), t_e(i)]$. Let $\mathscr{M}_f(i) \subseteq \mathscr{M}$ be the set of alternative materials that can be used for task i, so:

$$\begin{aligned}
\sum_{m\in\mathscr{M}(i)} \sum_{t=t_s(i)}^{t_e(i)} x_{h(i),i}^{mt} &= d_i\ i\in\mathscr{F}, && (5)\\
x_{h(i),i}^{mt} &= 0 \quad i\in\mathscr{F}, m\in\mathscr{M}\setminus\mathscr{M}_f(i), t\in\mathscr{T}, && (6)\\
x_{h(i),i}^{mt} &= 0 \quad i\in\mathscr{F}, m\in\mathscr{M}, t\in\mathscr{T} : t < t_s(i) \text{ or } t > t_e(i). && (7)
\end{aligned}$$

Let $\mathscr{X}$ denote the set of *excavation tasks*. Each task $i \in \mathscr{X}$ is associated to location $h(i)$, must be performed within the time window $[t_s(i), t_e(i)]$, and has an overall request d_i made by a set of materials $\mathscr{M}_x(i) \subseteq \mathscr{M}$. Coefficients φ_i^m (with $\sum_{m\in\mathscr{M}_x(i)} \varphi_i^m = 1$) determine the exact mix of the materials, so:

$$\sum_{m\in\mathscr{M}_x(i)} \varphi_i^m \sum_{t=t_s(i)}^{t_e(i)} x_{i,h(i)}^{m,t} = d_i\ \ i\in\mathscr{X}, \tag{8}$$

$$x_{i,h(j)}^{mt} = 0\ \ i\in\mathscr{X}, m\in\mathscr{M}\setminus\mathscr{M}_x(i), t\in\mathscr{T}, \tag{9}$$

$$x_{i,h(j)}^{m,t} = 0\ \ i\in\mathscr{X}, m\in\mathscr{M}_x(i), t\in T : t < t_s(k) \text{ or } t > t_e(k). \tag{10}$$

Temporary depots, dump sites, and quarries capacity constraints are as follows. Let $\mathscr{M}_s$ be the set of materials that can be stored into temporary depots and Q_k the maximum storing capacity for depot $k \in \mathscr{P}$, for each period. We have:

$$\sum_{m\in\mathscr{M}_s} f_k^{mt} \le Q_k\ \ k\in\mathscr{P}, t\in\mathscr{T}. \tag{11}$$

For each material $m \in \mathscr{M}_s$ and depot k let f_k^{m0} be the initial stock at the beginning of the first period, and UB_m the upper bound on the stock at the end of the last period.

$$\begin{aligned}
\sum_{(i,k)\in\mathscr{A}} x_{ik}^{mt} + f_k^{m,t-1} &= \sum_{(k,i)\in\mathscr{A}} x_{ki}^{mt} + f_k^{mt}\ \ k\in\mathscr{P}, m\in\mathscr{M}_s, t\in\mathscr{T}, && (12)\\
f_k^{m,|T|} &\le UB_m \qquad k\in\mathscr{P}, m\in\mathscr{M}_s. && (13)
\end{aligned}$$

Let C_k and Q_k^m be, respectively, the total capacity and the maximum capacity for each period and material, of a dump site $k \in \mathscr{D}$ or a quarry $k \in \mathscr{Q}$. Then:

$$\sum_{(i,k)\in\mathscr{A}} \sum_{t\in\mathscr{T}} \sum_{m\in\mathscr{M}} x_{ik}^{mt} \leq C_k \quad k \in \mathscr{D} \cup \mathscr{Q}, \tag{14}$$

$$\sum_{(i,k)\in\mathscr{A}} x_{ik}^{mt} \leq Q_k^m \qquad k \in \mathscr{D} \cup \mathscr{Q}, m \in \mathscr{M}, t \in \mathscr{T}. \tag{15}$$

Recycling constraints must be added to complete the model. For sake of simplicity we do not give explicitly these formulas, but only some hits on their nature (for a complete description the reader is referred to [7]). The recycling process is made by a first step where the High quality material from excavation is separated into different materials (in a separation plant), and by a second step where some of these materials are combined with bitumen and/or cement to obtain asphalt and concrete (in mixing plants). Each separation plant has its own and specified recipe, that states the quantities of exiting materials obtained from a unit of High quality material. Similarly each mixing plant has it own recipe to be used to obtain the required construction material. Finally capacity constraints exist for each plant and period.

Disaggregate formulation

Once the aggregate formulation has been used and a feasible solution is obtained (possibly after some iterations where the production department changes some quantities or constraints), the disaggregate model is used to define the single trips needed to execute all the material movements. The trip plan is made for all the time horizon, but any trip that begins in a time period must terminate in the same period. Moreover, each material must use a different type of truck, thus we can separate the model into several submodels, one for each period and material. From the aggregate solution we know, for each submodel, all origins and destinations for a given material. We use a set partitioning model where each column represent a possible origin-destination path, whose cost is normally non-linear, due to the complex regulations existing in the contracts with the transport suppliers.

6 The Decision Support System

The mathematical models of the previous section have been embedded into a decision support system. The process to use the DSS is highlighted in Fig. 4. It starts from the information of the Master Plan, which can be exported from Microsoft Project to Excel files. The description of the construction site network and of the external road network are also available as input files in Keyhole Markup Language. A Visual Basic application processes the master plan data and the network data, producing plain text input files for the aggregate model. The mathematical models are implemented and solved using the FICO Xpress Optimization Suite and the results are written on text files. The production department is responsible of the overall process, assisted by the STRAsco personnel. To facilitate the analysis of the solutions we have implemented two visualization tools, for the aggregate and disaggregate phases, respectively. The first tool is written in JavaScript and used the D3 library for visualizing data with HTML. The tool produces several diagrams showing the use of resources in all material locations and plants. For example, Fig. 5 reports on

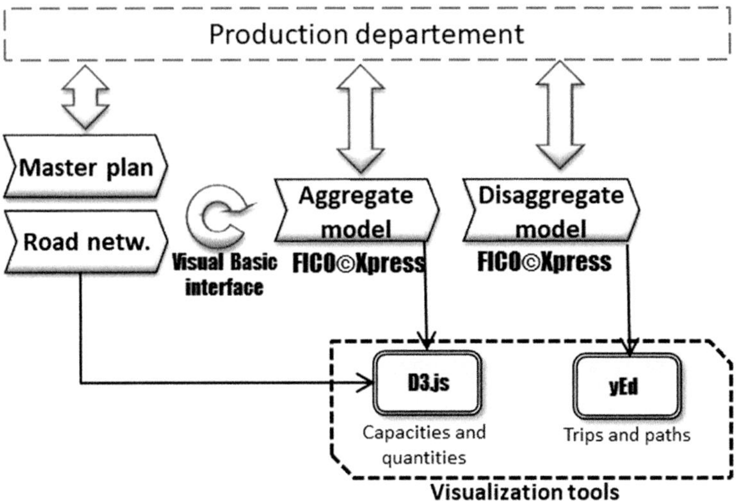

Fig. 4 The decision support system process

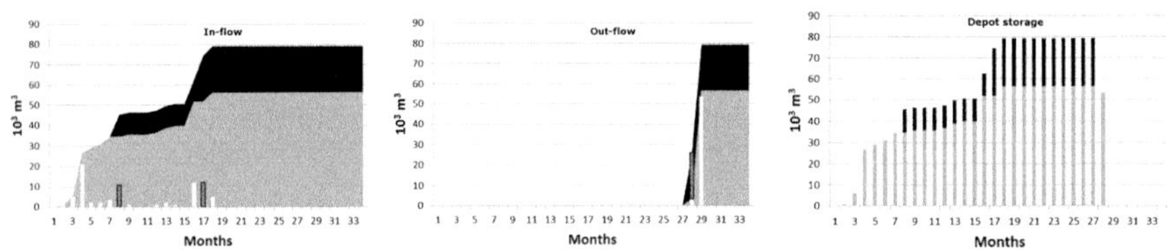

Fig. 5 In-out material flows and total storage in a temporary depot (area: cumulative quantities; histogram: monthly quantities)

the monthly flows entering end outgoing from a temporary depot, so as on the total storage. The second visualization tool gives detailed information on the planned trips producing maps of the paths as .graphml files that can be visualized and edited with the yEd editor. The Xpress optimization suite is the only commercial software used.

The DSS resulted to be an important instrument for the production department that drastically decreased the time needed for decision making and provides more detailed information than those that can be obtained with a manual solution. This allows to manage properly infeasibilities or uncommon peaks in costs. Due to the better knowledge of the possible implementation of the master plan, the definition of agreements with the external facilities was easier and more efficient, making possible to reduce under-used facilities and related costs. Finally, due to the optimization algorithm, the solutions provide better results that the manual one.

Without our DSS the production department took decisions on the usage of quarries, dump sites, renting of land for the temporary depots, vehicles usage and paths, etc., using a complete manual approach. They started from the master plan (Gantt chart) and using several Excel files for data management and temporary computations, tried to balance the flows of the materials while looking for a cheap solution. Indeed, the quantity of data involved, the number of constraints and the number of possible scenarios makes it very difficult even to find a solutions, so the department was usually concentrated to look for a feasible solution and only in a second step tried to adjust the solution to reduce the costs. Moreover, without the DSS, they were not able to look ahead for the complete time horizon, but for a small number of weeks. The manual approach requires several days for each complete solution.

After some preliminary test Strabag decided to manage the Autostrada Pedemontana Lombarda project using our DSS and dismissing completely the manual solutions.

7 Conclusions

In this paper we described a Decision Support System developed to assist the production department of Strabag AG in the management of the earthworks in a very large and complex construction project, namely the building of an Italian highway in the north of Milan, called Autostrada Pedemontana Lombarda. The DSS takes as input the master plan that the company commonly produces to manage any project. Using in cascade two mathematical programming models, solved with a commercial software, the DSS gives several advantages, with respect to the manual solutions previously adopted.

First of all a manual solution requires several days to be fixed, while the DSS runs in hundreds of seconds. It allows immediate detection and response to the infeasibilities of the master plan, which are very frequent due to the large number of resources, constraints, and time periods. By using the DSS in an interactive way, the production department can adjust the timing of the activities and the capacity of the resources, thus obtaining a feasible solution in a short time. Visualization tools help the engineers of Strabag to easily catch the main shortage and problems in an infeasible solution and to provide corresponding updating of the plan. Moreover, the DSS produces a detailed solution which is very useful to help in the definition of agreements with the external facilities (dump sites, quarries, transport operators, etc.), allowing to decrease unused facilities and related costs. Finally, the DSS optimizes the overall costs giving a direct economical benefit. The DSS implemented is very flexible and can manage linear costs as well as non-linear ones, that often appears when applying the complex rules of transport contracts.

Satisfied with this project, Strabag decided to use it on similar activities, without any further need of intervention by the operations research group of the University of Modena and Reggio Emilia. Strabag is currently discussing internally to embed it as a part of the Building Information Modeling system that is currently used to design and manage any medium/large project.

Acknowledgements Figures 1, 2 and 5 reprinted by permission, M. DellAmico, G. Fuellerer, G. Hfinger, M. Iori, S. Novellani, A Decision Support System for Highway Construction: The

References

1. Ahuja, R., Magnanti, T., Orlin, J.B.: Network Flows: Theory, Algorithms, and Applications. Prentice-Hall, Inc., Upper Saddle River, NJ, USA (1993)
2. Askew, W., Al-Jibouri, S.H., Mawdesley, M., Patterson, D.: Planning linear construction projects: automated method for the generation of earthwork activities. Autom. Constr. **11**(6), 643–653 (2002)
3. Aziz, R., Aboel-Magd, Y.: Suitably selection for earthwork equipment in Egyptian sites. Int. J. Educ. Res. **3**(1), 539–549 (2015)
4. Bogenberger, C., Dell'Amico, M., Fuellerer, G., Hoefinger, G., Iori, M., Novellani, S., Panicucci, B.: A two-phase earthwork optimization model for highway construction. J. Constr. Eng. Manag. **141**(6), 05015003 (2015)
5. Burdett, R., Kozan, E.: An integrated approach for earthwork allocation, sequencing and routing. Eur. J. Oper. Res. **238**(3), 741–759 (2014)
6. Burdett, R., Kozan, E., Kenley, R.: Block models for improved earthwork allocation planning in linear infrastructure construction. Eng. Optim. **47**(3), 347–369 (2015)
7. Dell'Amico, M., Fuellerer, G., Hoefinger, G., Iori, M., Novellani, S.: A decision support system for highway construction and its application to the Autostrada Pedemontana Lombarda. Interfaces **46**(3), 203–280 (2016)
8. Easa, S.M.: Earthwork allocations with nonconstant unit costs. J. Constr. Eng. Manag. **113**(1), 34–50 (1987)
9. Easa, S.M.: Earthwork allocations with linear unit costs. J. Constr. Eng. Manag. **114**(4), 641–655 (1988)
10. Hare, W.L., Koch, V.R., Lucet, Y.: Models and algorithms to improve earthwork operations in road design using mixed integer linear programming. Eur. J. Oper. Res. **215**(2), 470–480 (2011)
11. Karimi, S., Mousavi, S.J., Kaveh, A., Afshar, A.: Fuzzy optimization model for earth-work allocations with imprecise parameters. J. Constr. Eng. Manag. **133**(2), 181–190 (2007)
12. Marzouk, M., Moselhi, O.: Multiobjective optimization of earthmoving operations. J. Constr. Eng. Manag. **130**(1), 105–113 (2004)
13. Moselhi, O., Alshibani, A.: Optimization of earthmoving operations in heavy civil engineering projects. J. Constr. Eng. Manag. **135**(10), 948–954 (2009)
14. Nassar, K., Hosny, O.: Solving the least-cost route cut and fill sequencing problem using particle swarm. J. Constr. Eng. Manag. **138**(8), 931–942 (2011)
15. Parente, M., Cortez, P., Correia, A.G.: An evolutionary multi-objective optimization system for earthworks. Expert Syst. Appl. **42**(19), 6674–6685 (2015)
16. Rahmani, A.M., Walewski, J.: Two-dimensional earthwork optimization using mixed-integer programming. In: Transportation Research Board, 90th Annual Meeting, Washington, DC, No. 42(19), 11-1975 (2015)
17. Tam, C.M., Tong, T.K.L., Wong, B.W.L.: An integrated system for earthmoving planning. Constr. Manag. Econ. **25**(11), 1127–1137 (2015)
18. Zhang, H.: Multi-objective simulation-optimization for earthmoving operations. Autom. Constr. **18**(1), 79–86 (2008)

A Comparison of Optimization Models to Evaluate the Impact of Fuel Costs When Designing New Cruise Itineraries

Daniela Ambrosino, Veronica Asta and Federico Bartoli

Abstract The itinerary design is a problem that belongs to the class of cruise supply. The development of cruises' itineraries to offer on the market is a long process. Given a ship located in a specific world's basin, given a duration and a homeport, an itinerary has to be planned choosing a sequence of ports to visit among a set of available ones. The itinerary must be characterized by a schedule, which comprehends the arrival time and the departure time of the ship for each port. The objectives to pursue in the decisions are the maximization of both the customer satisfaction and the revenue and the minimization of the costs. In this paper the focus is on costs, that depend on fuel consumption and ports' services. The fuel cost is function of the speed of the ship. Starting from a model recently proposed in the literature for solving the cruise itinerary designing problem (CIDP) (Ambrosino et al. in Proceedings of 15th IFAC Symposium on Control in Transportation Systems, Savona, Italy, June 2018, [3]) in which fuel costs were based on classes of speed, a different model has been developed for including a more precise fuel cost computation. The fuel costs represent a great part of the company costs, influencing the operative margin. Thus, the fuel consumption is here computed in terms of metric tons for each single speed. The model has been tested solving real cases of Costa Crociere in the Mediterraneo basin. The obtained solutions have been compared, both in terms of cruise route and objectives values with those obtained by the previous model used by the company to define optimal itineraries.

Keywords Cruise itinerary design · Optimization model

D. Ambrosino (✉)
DIEC, Universitá di Genova, Genova, Italy
e-mail: daniela.ambrosino@economia.unige.it

V. Asta
CIELI, Universitá di Genova, Genova, Italy
e-mail: veronica.asta@economia.unige.it

F. Bartoli
Costa Crociere Spa, Genova, Italy
e-mail: federico.bartoli@costa.it

M. Dell'Amico et al. (eds.), *A View of Operations Research Applications in Italy, 2018*, AIRO Springer Series 2, https://doi.org/10.1007/978-3-030-25842-9_14

1 Introduction and Literature Review

The worldwide cruise ship tourism just account for about 2% of the world tourism market's revenue, of approximately 35.5 billion U.S. dollars in 2016 [1] and it is forecasted to growth in the next years. Also, the passenger capacity of 466 thousand in 2016 is expected to grow for reaching 521 thousand by 2018. The cruise customers are usually loyal/repeat clients, and each cruise company wishes to offer new attractive and desirable itineraries in order to satisfy its loyal customers and maximise its revenue. For these reasons, each itinerary is a crucial element of the cruise offer. In a recent work [5] the authors recognize that the itinerary represents an important task having an impact on both consumers' satisfaction and local economies; the role of the different ports (and cities) within different itineraries has been analyzed. In [10] is stressed the role of the itineraries in the appeal of a cruise service. In [9] the effect of cruise tourism (in particular, the impact of excursion packages offered to the passengers of a cruise) on the local communities has been investigated. The development of cruises' itineraries to offer on the market is a long process involving different planning levels [2]. The itinerary design problem belongs to the lower decision level (the day by day planning) and consists in determining the exact route of a ship in a basin. It is influenced by decisions taken at the previous levels, such as the duration of the trip and the homeport (i.e. the main port of departure corresponding to the last port of the cruise, generally used for the major part of the procurements). Thus, an itinerary is a sequence of ports visited by a ship with a particular schedule, that is an arrival time and a departure time for each port. At each port customers usually leave the ship in such a way to visit the city and the surrounds area. While defining a new itinerary the main aims are the maximization of both the customer satisfaction and the revenue and the minimization of the costs. The choice of the ports to include in the itinerary has a great impact on both the customer satisfaction and the revenue. On the other hand, the ports included in the route have also an impact on the costs represented, in particular, by ports' services and fuel consumption [6]. The port's costs are fixed amounts paid when the port is visited. The fuel cost is function of the speed of the ship. Thus, the distance between couple of ports visited by the ship and the schedule have an impact on the speed of sailing and thus on the fuel consumption. The schedule has also an impact on the revenues: the revenue arises from the sale of both the excursions tickets and the on-board services, and depends on the length of the ports' stop and, as a consequence, on the hours on sailing. In [4] the authors stress that a cruise line's duration of stay in a port is influenced by many factors, among others, the sailing distance from the previous port, the sailing distance to the next port, the nature of international cruise lines, the specific home ports of cruise ships and the attractive ports of call.

This work, realized in collaboration with Costa Crociere, a cruise company operating in Europe, focuses on the day by day planning having as main aim the definition of a new itinerary. In particular, the present analysis is based on the evaluation of the impact of fuel cost on the definition of a new itinerary. Costa Crociere is a branch of Costa Group, which belongs to Carnival Corporation, the largest holding in the industry. Costa Crociere offers its products, the cruises services, to many markets. The main markets on which it works with are Italy, France, Spain, Switzerland and Austria. The decisional process followed by the company for identifying a new itinerary has been deeply analysed in [2]. In accordance with the company main requirements, in [3] a mathematical formulation for determining a new itinerary maximising revenues, customers satisfaction, while minimizing operative costs has been developed. The obtained itineraries have been compared with the existing ones permitting the company to perform some analysis on the efficiency of their itineraries. In [7] the authors propose a model for the tour planning problem that is for defining a fixed number of tours, having multiple time windows and the possibility to choose among different travel speed values. Their model is a variant of the vehicle routing problem.

In the present work, the model proposed in [3], which was based on both slots of half and full days for defining the ship stop duration at ports, and fuel costs based on classes of speed, has been extended for including a more precise fuel cost computation. The fuel costs represent a great part of the company costs, influencing the operative margin. Thus, the fuel consumption is here computed in terms of metric tons for each single speed instead of classes of speed. In particular, in [3] two classes of speed are considered: the first class comprehends speeds for high consumption, i.e. until sixteen and exceeding seventeen nodes, and the second one includes speeds for low consumption, i.e. between sixteen and seventeen nodes. In the current paper it is known the fuel consumption for each speed, from eight to eighteen nodes.

The proposed model has been tested solving real cases in the Mediterraneo basin and used to evaluate the impact of the fuel cost on cruise offer.

The importance of the consumption control is stressed in [11] where the authors also analyse the effect of the emission control areas (ECA) on cruise shipping. The authors proposed a mixed integer programming model to re-schedule voyage plans by optimizing speeds, sailing patterns and ports of call sequences. This work is different from [11] since the model proposed is used for the definition of the cruise itinerary while minimizing fuel costs and includes some requirements related to the characteristic that the itinerary must have.

The paper is organized as follows: the problem under investigation is detailed in Sect. 2, together with the solution approach. The case study is reported in Sect. 3 and some conclusions are reported in Sect. 4.

2 The Problem Under Investigation and the Solution Approach

Given a ship located in a specific world's basin, given a duration and a homeport, an itinerary has to be planned choosing a sequence of ports to visit among a set of available ones. How to choose the ports to include in the itinerary is the difficult question faced by the itinerary planners of the cruise industry. The main goals are the following:

- the customer satisfaction: passengers may decide to repeat a cruise and can recommend it to their friends, thus the accessibility and the appealing of the destinations achievable from the port are really important;
- the revenue arising from the sale of both the excursions tickets and the on board services;
- the costs for offering the cruise must be monitored and a positive operative margin must be obtained for being competitive in a complex market: the average revenue for passenger is about 1,779 dollars, with expenses per passenger very high, that bring an average profit of only 226 dollars (data related to 2015 [1]).

Appealing and accessibility values linked to the different ports have been used as customer satisfaction measure. The company has evaluated these values by analyzing past cruises results, passengers opinions, marketing analysis. Note that some factors like the length of the port's stop and the season in which the cruise is offered can influence these values. Moreover, the value associated to each port for reflecting the customer satisfaction varies depending on both the other ports visited by the ship during the journey, and the number of ports included in the itinerary.

The revenues included in the current analysis are those arising from the sale of both the excursions tickets and the on board services. The on board services can be distinguished in services offered during sailing, when the ship is docked at a port, and services offered only during sailing some miles away from the coast (i.e. the Casino). All these revenues depend on the length of the ports' stop and on the hours on sailing.

In this work, as in [3], each day of the cruise journey is split into four daily slots: morning, afternoon, evening and night. It is assumed that the ports' stop can be half day equivalent to six hours (i.e. morning, afternoon or evening) or full day equivalent to twelve hours (two continuous slots, i.e. morning and afternoon or afternoon and evening), and the cruise cannot stop at the port during the night. Moreover, some requirements related to the arrival time at each port are imposed differently for half day and full day stops.

The costs included in the current analysis are fuel cost and cost for ports' services. The fuel cost depends on both the fuel consumption and the fuel price. The fuel consumption is known and expressed in terms of metric tons and it is function of the speed of the ship. Obviously, the speed and, consequentially, the consumption depend on the distances among the visited ports and on the length of the ports' stop. The port's costs are fixed amounts paid only if the related port is selected.

Note that, each itinerary must be characterized by a selection of ports belonging to at least two different countries; otherwise, it becomes a ferry service and the tax system changes. Moreover, the guests safety is very important and drives the ports selection: the ship must only visit the areas with no political issues or risks (geopolitical safety), and only the ports or the facilities that have a series of specific standard, for granting a safety dock (facility's safety). Geopolitical problems are really time and space depending, thus it is important to include this specific characteristic of the ports in the input data of the model. Finally, the company may fix a limit on the number of ports to select for the itinerary, or may decide to limit the number of ports having the same characteristics, i.e., beach or culture ports.

2.1 The Solution Approach

In the present subsection the proposed approach for solving the problem described above is introduced. In particular, being this approach based on a mathematical model derived by the one proposed in [3], only the main novelty will be discussed in details, while the other parts of the model will be only briefly described. The model proposed in [3] was the first attempt to solve the itinerary problem for Costa Crociere. The novelty here included is related to a more accurate computation of the fuel cost: for each speed from eight to eighteen nodes, it is known the fuel consumption and thus, the speed of the whole itinerary is optimized in accordance to these fuel data.

Given M a set of days for planning the cruise, and a set P of available ports, the problem consists in defining the sequence of ports to visit, specifying the day, the time of arrival and the time of departure for each port included in the itinerary, and the speed used for travelling from one port to another one. The main aims included in the objective function of the mathematical model are the following.

- The maximization of:

1. accessibility, computed as the sum of the accessibility value of each selected port ($ac_{p,f}$), that depends on the stop duration f (i.e., half day/full day);
2. appealing, considering the whole set of ports belonging to the itinerary and thus influencing the total appealing value. In fact, this latter is computed as the sum of the appealing value of each selected port ($ap_{p,f}$), that depends on the stop duration (i.e., half day/full day) plus a correction ($cv_{p,q}$) that permits to link the customer satisfaction to the whole trip analyzing when two ports p and q are included in the route, even if they are not visited consecutively;
3. revenue deriving from the sale of excursions ($er_{p,f}$) and from the sale of on board services, distinguishing between services sold on board when the ship is in a port ($or_{p,f}$) and during navigation ($sr_{p,q,f}$). All the revenues depend on the port's stop and the trip length (f).

- The minimization of:

1. ports' costs, given by the sum of the fixed service cost (c_p) of the selected ports;
2. fuel costs obtained as the sum of the fuel costs supported by the company for connecting each couple of ports visited by the ship. The fuel cost for connecting port p and q depends on the speed of the ship, knowing: the distance between ports p and q ($d_{p,q}$), the speed (sp_o) used as explained later (see constraints (3)), the fuel consumption for nautical mile if the speed is o (fc_o) and the fuel price per metric ton (fp). Note that a minimum and a maximum speed of the ship (s^{min}, s^{max}) are known.

In the objective function some weights for the appealing and the accessibility values have been used, in accordance with the feelings of the company's managers.

The itinerary is a sequence of ports visited by the ship and must have some characteristics specified by the cruise company, such as:

- must start from a specified home-port and finish at the same port;
- a given number of ports, in accordance with the limits imposed by the company (p^{min}, p^{max}), must be visited;
- a maximum number of ports belonging to the same cluster can be selected, having defined C the set of clusters used for grouping ports in accordance with their characteristics (i.e. beach, culture, etc.);
- must avoid the unsafe ports, in accordance with an evaluation updated continuously;
- must have ports belonging to more than one country, having defined H the set of countries to which the ports belong to;
- must guarantee a minimum level of accessibility (ac^{min}) for customers.

Moreover, some constraints are imposed for the ports' arrival time, departure time and stop duration:

- the ship starts its trip leaving the homeport at time 0 and must come back to the homeport within a given time limit ($Maxt$);
- the duration of the ports' stops (z_p in hours) must be defined as half day (six hours) or full day (twelve hours);
- each half day port's stop must belong to one slot among the available ones in K, that is the set of slots composing each day of the cruise, i.e. morning, afternoon, evening and night;
- each full day port's stop must belong to two consecutive slots (as before among the available ones in K);
- the ship can have no more than one first dock (stop) for each port and can have no more than one first dock for each day;
- the arrival time at each selected port must be within a given time window: ($a^E_{k,m}$, $a^L_{k,m}$) respectively the earlier and the latest arrival time (e.g. from midday and until two o'clock p.m. if the port's stop is half day in the afternoon slot).

On the other hand, it is important to define the time at sea: the time spent at sea between two consecutive ports p and q, that is the hours on sailing between p and q, is defined in accordance with the departure time from port p (d_p) and the arrival time in port q (a_q). The hours on sailing between the ports called consecutively must be within an upper and a lower limit (h^{max}_{pq}, h^{min}_{pq}). We define hours on sailing as half or full day depending on the duration, less than or equal to six hours and greater than six hours.

We can observe that, the time spent on sailing impacts on both the on board revenue during navigation and the fuel costs. In fact, from the time spent in navigation between ports p and q, and the distance between port p and q ($d_{p,q}$), depend the speed of sailing between p and q (i.e. variables $v_{p,q}$). Note that, $v_{p,q}$ is limited by an upper bound and a lower bound defined in (4) and (5), which depend on both the upper bound limits on the hours of sailing between the ports and the minimum and maximum speed of the ship.

Assuming for the consumption a step-wise function of the speed, we have defined set O of the speeds whose elements sp_o are used for both defining the speed of sailing $v_{p,q}$ and fixing to one the binary variables ($l_{p,q,o}$) used for computing the fuel costs. Fuel costs are computed thanks to Eq. (1).

$$\sum_{p\in P}\sum_{q\in P}\sum_{o\in O} fc_o l_{p,q,o} d_{p,q} fp \tag{1}$$

Summarizing, the new constraints to include in model [3] for computing the fuel consumption and relative costs, are the following:

$$\sum_{o\in O} l_{p,q,o} = g_{p,q} \quad \forall p, q \in P \tag{2}$$

$$v_{p,q} = \sum_{o\in O} sp_o l_{p,q,o} \quad \forall p, q \in P \tag{3}$$

Constraints (2) impose to choose only one speed for going from port p to port q when p and q are called sequentially (i.e., $g_{p,q} = 1$). Thanks to (3), and the following couple of constraints (4, 5), it is possible to define $v_{p,q}$. Note that (4) and (5) are derived from the equation of straight line through two points, knowing both the maximum speed of sailing related to the minimum hours needed to sail between ports p and q, and the minimum speed of sailing related to the maximum hours needed to sail between ports p and q.

$$\frac{\left(s^{min} - s^{max}\right)\left(s_{p,q} - h_{pq}^{min}\right) + s^{max}\left(h_{pq}^{max} - h_{pq}^{min}\right) g_{p,q}}{h_{pq}^{max} - h_{pq}^{min}} - 0,5 \le v_{p,q} \quad \forall p, q \in P \tag{4}$$

$$\frac{\left(s^{min} - s^{max}\right)\left(s_{p,q} - h_{pq}^{min}\right) + s^{max}\left(h_{pq}^{max} - h_{pq}^{min}\right) g_{p,q}}{h_{pq}^{max} - h_{pq}^{min}} + 0,5 \ge v_{p,q} \quad \forall p, q \in P \tag{5}$$

3 The Case Study and the Relative Results

The new model obtained by combining the model proposed in [3] and Eqs. (2)–(5) has been used to solve a real case of Costa Crociere, that is to design a new cruise itinerary in the West Mediterranean area for each season. These new itineraries have been compared with both the itineraries obtained with the first model developed for the Cruise Itinerary Designing Problem (CIDP) in [3], and an itinerary operated in the last years by a ship of the fleet of Costa Crociere, that is Costa Diadema. The itineraries that must be defined is characterized by Savona as homeport and a duration of 7 nights. The ports of the West Mediterranean area included in the analysis are the following: Savona, Ajaccio, Toulon, Marseille, Barcelona, Mahon, Palma, Ibiza, Valencia, La Goulette, Valletta, Palermo, Catania, Cagliari, Naples, Civitavecchia, Livorno, La Spezia. These ports have been grouped into five clusters, each one with determined characteristics: Italian culture, France culture, Spanish culture, Seaside destination and Other. Within this ports, at the moment the port of La Goulette, located in Tunisia, is unsafe, and thus not selectable. The number of selectable ports to include in the new itineraries can be from four to six. The speed of sailing of the ship has been limited in a range from eight to eighteen nodes. The level of accessibility to grant for each itinerary must be bigger than a value fixed in accordance with the itinerary planners experience of Costa Crociere.

As far as the arrival times are considered, the following limitations have been imposed: dock in the morning—arrival time from eight o'clock to ten o'clock a.m.; dock in the afternoon—arrival time from two o'clock to four o'clock p.m. Weights α and β have been fixed in accordance with the itinerary planners of the company.

The proposed model has been implemented in Mathematical Programming Language (MPL) and solved by the commercial solver Gurobi 7.0.2. on a PC Intel Core i3, 2.00 GHz; 4G RAM. The solved models have 6629 variables and 6005 constraints. The optimal solutions have been obtained in 10 s on average.

The obtained results are shown in the following figures.

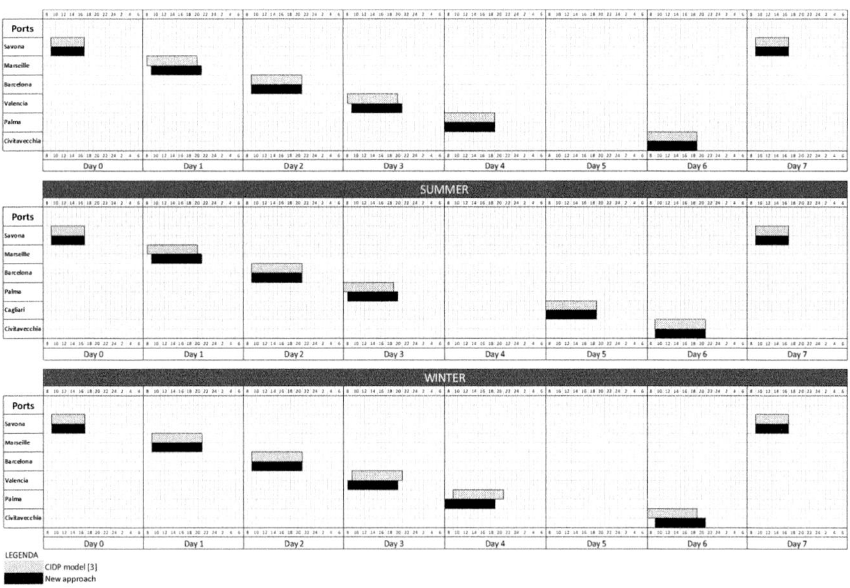

Fig. 1 Itineraries comparison

	Objective	Appealing	Fuel cost
SHOULDER	0.67%	0.00%	-4.39%
SUMMER	7.44%	7.57%	-8.36%
WINTER	5.74%	6.70%	-4.59%

Fig. 2 Percentage deviation for comparing the proposed solution approach and the CIDP model [3]

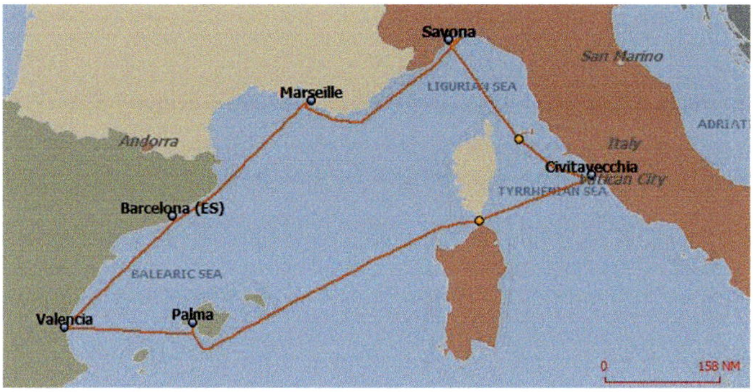

Fig. 3 Shoulder and winter itinerary

Looking at Fig. 1 it is possible to note that the itineraries defined in each season are exactly the same in the two cases, excepted for the arrival and departure time at ports. The schedule of each itinerary proposes six ports with full day stops. A full day at sea is scheduled in day 5, except for the summer itinerary that presents the day 4 on sailing. The fuel cost function included in the new model has an impact on the definition of the hourly schedule of the itineraries, on costs and operative margin. As stressed in [8], there is a seasonality in the fuel consumption and emission, and thus the impact of the fuel cost in the itinerary is greater in summer, and very low in springer and autumn (here denoted shoulder).

Figure 2 shows, for the different seasons, the percentage deviation between the values of objective function, appealing and fuel cost obtained by using the here described solution approach and the model in [3]. It is evident that, when using a function representing more precisely the fuel costs, in winter and summer higher values of appealing and lower costs of fuel permit to gain in terms of objective function. The higher saving in the fuel costs is in summer.

Finally, Fig. 3 shows the shoulder and winter itinerary and Fig. 4 the summer one, that corresponds to the Costa Diadema itinerary during Summer 2016 and Summer 2017.

The obtained itineraries are very similar to those offered by the company in the West Mediterranean basin (see Figs. 3, 4 and 5). Anyway, we have noted that, as far as the full day on sailing is considered, in these solutions it is always in the middle of the route (day 4 or more frequently day 5). The company generally prefers to include either an initial full day at sea (could increase the excursions revenue) or a final full day at sea (could increase the on board revenue).

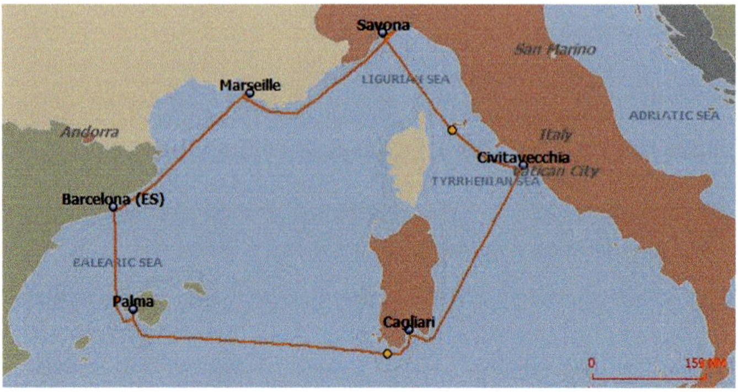

Fig. 4 Summer itinerary and Costa Diadema during summer 2016 and summer 2017

Fig. 5 Costa Diadema during the following seasons: autumn, winter and spring 2015/2016, autumn, winter and spring 2016/2017, autumn and winter 2017/2018

Looking at Figs. 3, 4 and 5 the only difference is the call at Valencia instead of the call at La Spezia and, as a consequence, the full day at sea is day 4 instead of day 5. The objective function's values of the compared itineraries have been analyzed. They cannot be reported for privacy issues. The Costa Diadema's itinerary has bigger excursions revenues but lower on board revenues. The same situation concerns costs: the optimal itinerary found has a huge fuel savings even if it has a bigger amount of ports' costs. The appeal is slightly better for the cruise route already offer from the company but for the accessibility value is the opposite.

Analyzing these differences among the itineraries already offered by the company and the new ones defined by using the proposed solution approach (see Sect. 2.1) we can conclude that the day of navigation in the middle of the route permits to save more fuel costs thanks to an efficient speed during the route.

Just to conclude, tables reported in Fig. 6 compare, for the itineraries of each season, the speed of sailing for approaching the ports and total fuel costs obtained by the model proposed in [3] and thanks to the new approach.

SHOULDER

Routes		Speed	
From	To	CIDP model [3]	New approach
Savona	Marseille	17.00	16.00
Marseille	Barcelona	17.00	17.00
Barcelona	Valencia	16.36	16.00
Valencia	Palma	14.03	16.00
Palma	Civitavecchia	16.00	16.00
Civitavecchia	Savona	16.72	17.00

SUMMER

Routes		Speed	
From	To	CIDP model [3]	New approach
Savona	Marseille	17.00	16.00
Marseille	Barcelona	16.27	17.00
Barcelona	Palma	16.00	16.00
Palma	Cagliari	12.31	13.00
Cagliari	Civitavecchia	18.00	18.00
Civitavecchia	Savona	17.33	17.00

WINTER

Routes		Speed	
From	To	CIDP model [3]	New approach
Savona	Marseille	16.12	16.00
Marseille	Barcelona	17.00	17.00
Barcelona	Valencia	16.12	16.00
Valencia	Palma	15.62	16.00
Palma	Civitavecchia	16.00	16.00
Civitavecchia	Savona	17.00	17.00

Fig. 6 A comparison of the speed of sailing of the itineraries obtained by using the proposed solution approach and the CIDP model proposed in [3] for the different seasons

In all cases the total fuel cost has been reduced thanks to the new approach. This reduction is about 4.5% for winter and shoulder and 8.4% for summer itineraries; in absolute terms it seems very little but this is only one itinerary in a week, summarizing for all itineraries and for each year this reduction can represent a very important value for the company.

4 Conclusions and Future Works

This work proposes an extension of a model presented in the recent literature for defining new itineraries, in order to be able to represent in a more accurate way the consumption and thus the fuel costs of the cruise company. The first attempt to define optimal cruise itineraries [3] was an optimization model in which fuel costs were based on classes of speed. The comparison among the obtained itineraries furnishes some interesting results. Some further analysis will be conduct in order to realize if some more savings in fuel costs can be realized operating with a more flexible timetable of the slot stops.

Since having a set of efficient solutions among to choose can efficiently support the decision-making process of the company, future analysis should be related to multi-objective approaches for defining new itineraries. Anyway, also these first optimization approaches, very appreciated by the company, can be used in all basins operated by Costa Crociere for defining new appealing and cost savings itineraries.

References

1. www.statista.com
2. Ambrosino, D., Asta, V.: Evaluation of cruise itineraries offer. Int. J. Hosp. Manag. (submitted for publication)
3. Ambrosino, D., Asta, V., Bartoli, F.: An optimization model to design a new cruise itinerary: the case of Costa Crociere. In: Proceedings of 15th IFAC Symposium on Control in Transportation Systems, Savona, Italy, June 2018
4. Chen, J.M., Nijkamp, P.: Itinerary planning: modelling cruise lines lengths of stay in ports. Int. J. Hosp. Manag. **73**, 55–63 (2018)
5. Cusano, M.I., Ferrari, C., Tei, A.: Port hierarchy and concentration: insights from the Mediterranean cruise market. Int. J. Tour. Res. **19**(2), 235–245 (2017)
6. Fagherholt, K., Gausel, N., Rakke, J., Psaraftis, H.: Maritime routing and speed optimization with emission control areas. Transp. Res. Part C **52**, 57–73 (2015)
7. Mancini, S., Stecca, G.: A large neighborhood search based matheuristic for the tourist cruises itinerary planning. Comput. Ind. Eng. **122**, 140–148 (2015)
8. Maragkogianni, A., Papaefthimiou, S.: Evaluating the social costof cruise ships air emissionsin major ports of greece. Transp. Res. Part D **36**, 10–17 (2015)
9. Parola, F., Satta, G., Penco, L., Persico, L.: Destination satisfaction and cruiser behavior: the moderating effect of excursion package. Res. Transp. Bus. Manag. **3**, 53–64 (2014)
10. Rodriguea, J., Notteboomb, T.: The geography of cruises: itineraries, not destinations. Appl. Geogr. **38**, 31–42 (2013)
11. Zhen, L., Li, M., Hu, Z., Lu, W., Zhao, X.: The effects of emission control area regulations on cruise shipping. Transp. Res. Part D **62**, 47–63 (2018)

An Integrated Algorithm for Shift Scheduling Problems for Local Public Transport Companies

Claudio Ciancio, Demetrio Laganá, Roberto Musmanno and Francesco Santoro

Abstract This paper presents an integrated approach to solve two shift scheduling problems for local public bus companies: the first one aims at finding a schedule for vehicles, given a set of rides to do; the second one aims at assigning drivers to vehicle schedules. The first subproblem to be faced is the *Multiple Depot Vehicle Scheduling Problem*, while the second problem to deal with is the *Crew Scheduling Problem* where each trip is assigned to a driver. These problems are known to be NP-Hard. The main difference with respect to the literature is that the mathematical model, and the related algorithm, are designed based on real world-requirements. Computational results have been carried out on large real-word instances.

Keywords Vehicle scheduling · Crew scheduling · Simulated annealing · Local search

1 Introduction

Public transport companies usually face two main cost items: driver salaries and operational vehicle costs. Therefore the use of algorithms aimed at reducing the number of vehicles and drivers employed is a topic of great interest. Usually this problem is split into two subproblems where the output of the first becomes the input for the second. In the first problem, namely, *Multiple Depot Vehicle Scheduling Problem* (MDVSP) [3, 9], a set of routes performed by a set of vehicles located in different depots is determined. The constraints of this problem are related to the

---

An extended version of this article has been published in Omega The International Journal of Management Science, Vol 75, Author(s): Claudio Ciancio, Demetrio Laganá, Roberto Musmanno, Francesco Santoro, Title of article: An integrated algorithm for shift scheduling problems for local public transport companies, Pages No. 139–153, Copyright (2018), with permission from Elsevier.

C. Ciancio · D. Laganá (✉) · R. Musmanno · F. Santoro
Department of Mechanical, Energy and Management Engineering,
University of Calabria, 87036 Rende, CS, Italy
e-mail: demetrio.lagana@unical.it

M. Dell'Amico et al. (eds.), *A View of Operations Research Applications in Italy, 2018*,
AIRO Springer Series 2, https://doi.org/10.1007/978-3-030-25842-9_15

features of the available vehicles and the trips to execute. In the second problem, namely, *Crew Scheduling Problem* (CSP) [2, 5], each trip is assigned to one driver according to safety conditions and contractual restrictions. However, in many cases even if this approach allows to reduce significantly the computational time, the overall solution could be not satisfactory because it is obtained by solving separately two distinct problems. The aim of the proposed technique is to keep the benefits given by a sequential approach, while considering at the same time the link between the two problems. Furthermore these two problems are subject to several constraints drawn from EU legal framework (EU regulation no. 561/2006), while other constraints come from an analysis of the domain of application. This fact would make attractive the proposed algorithm as for industry applications. The paper is organized as follows: first the problem is introduced; taxonomy and the state of the art are presented. Then two subproblems in which the overall problem is decomposed are described, and a heuristic algorithm for each subproblem is proposed. Finally computational results are presented and discussed.

2 Problem Description

The service provided by a Public Transport Company is made by a set R of *transportation activities* performed by a set of *vehicles* that run within the boundaries of a geographic area. A vehicle is driven by a *driver* or, more generally, by a *crew*. A transportation activity is named *trip* (or *ride*) and it is characterized by a route that links two points in the geographic area. The geographic area is represented through a set of points of interest involved in the service provided by the company that can be classified as follows: *stop*, a point where passengers can board on or get off vehicles; *depot*, a point where vehicles start and end their daily shifts; *parking point*, a point where a vehicle can lay over for a long time; *refuelling station* a point where a vehicle can be refueled; *relief point* a point where a bus driver can change vehicle. For a Public Transport Company, the objective of the shift planning process is to execute all the trips that have been considered during the service design phase. To this aim the following resources must be exploited at best: *vehicles* and *crews*.

The shift planning process gives a set of daily shifts for the resources listed above as output. A shift ω is composed by a sequence A of activities sorted by time. For each activity $a \in A$ the following data are given: $p_s(a)$ and $p_e(a)$ are the points of interest where a starts from and ends to; while $t_s(a)$ and $t_e(a)$ are times when a begins and finishes at.

Two activities $a_1 \in S$ and $a_2 \in S$ are said to be *time continuous* (or merely *continuous*) if $t_e(a_1) = t_s(a_2)$. Two activities $a_1 \in S$ and $a_2 \in S$ are said to be *space contiguous* (or merely *contiguous*) if $p_e(a_1) = p_s(a_2)$. A sequence of activities $A = \{a_1, \ldots, a_k, \ldots, a_n\}$ performed by a resource (vehicle or driver) is *sound* if each couple of consecutive activities (a_k, a_{k+1}) is time continuous and space contiguous. A shift σ is sound if it is composed by a sequence A that is sound in turn.

Finally the objective of the shift planning process is to find a set of shifts that are sound and comply to all the constraints that are defined for the specific resource.

2.1 The Vehicle Scheduling Planning Phase

The vehicle scheduling planning phase aims at minimizing the number of used vehicles and minimizing the dead rides in terms of total mileage. This phase defines which trip of R must be assigned to which vehicle. The input data for the vehicle scheduling problem are:

- the set $V = \{v_1, \ldots, v_i, \ldots, v_u\}$ of vehicles available to perform the trips. For each $v_i \in V$ the following features are defined: its type τ_{v_i}; the amount k_{v_i} of kilometers that v_i can run without refuelling, a fixed cost K related to its use.
- the set $R = \{r_1, \ldots, r_i, \ldots, r_k\}$ of trips. For each $r_i \in R$ the following features are defined: $p_s(r_i)$ is the stop where r_i starts from; $p_e(r_i)$ is the stop where r_i ends to; $t_s(r_i)$ is the time when r_i begins at; $t_e(r_i)$ is the time when r_i finishes at; $l(r_i)$ is the length of r_i in km; the set $\mathscr{T}(r_i) = \{\tau_1, \ldots, \tau_i, \ldots, \tau_q\}$ of all the vehicle types that can be assigned to r_i.
- the set $S = \{s_1, \ldots, s_i, \ldots, s_w\}$ of stops; for each $s_i \in S$ the corresponding coordinates are defined.
- the set D of depots;
- the set $S_p \subseteq S$ of parking points;
- the set $S_r \subseteq S$ of refuelling stations.

We denote by $l^v_{i,j}$ and $t^v_{i,j}$ the distance and the time that vehicle v requires to move from the end location of trip r_i to the start location of trip r_j, respectively. These parameters depends on vehicle v because some bus stops can be used only by a particular type of vehicle. Therefore, the distance traveled to connect two trips could change according to the vehicle type used to perform it. The length $l^v_{i,j}$ of each transfer is computed by using the Manhattan distance. For a given vehicle v, a shift ω_v is composed by all the daily activities, named *rides*, assigned to v, which departs from a depot d (*pull-out*) and returns to. The activities that form a vehicle shift ω_v may be as follows: *ride run*, activities in which the vehicle is used to provide a service to the passengers; *dead run*, activities needed to move a vehicle across the geographic area without passenger transportation, the set of all the dead rides is $R' = \{r'_1, \ldots, r'_i, \ldots, r'_k\}:\{R \cap R' = \emptyset\}$. The dead run types are detailed in the following:

- *pull-out/pull-in*: each vehicle starts the service by traveling from the depot to the starting point of the first trip assigned and ends by traveling from the the end point of the last trip to the depot;
- *transfer*: for each pair of consecutive trips (r_i, r_j) served by a vehicle v, if $p_e(r_i) \neq p_s(r_j)$, a transfer is needed to move v from $p_e(r_i)$ to $p_s(r_j)$. There are two different type of transfers: *simple transfer* (if v can move from $p_e(r_i)$ to $p_s(r_j)$ and does not need to lay over a parking point or, equally, parking is permitted at least in

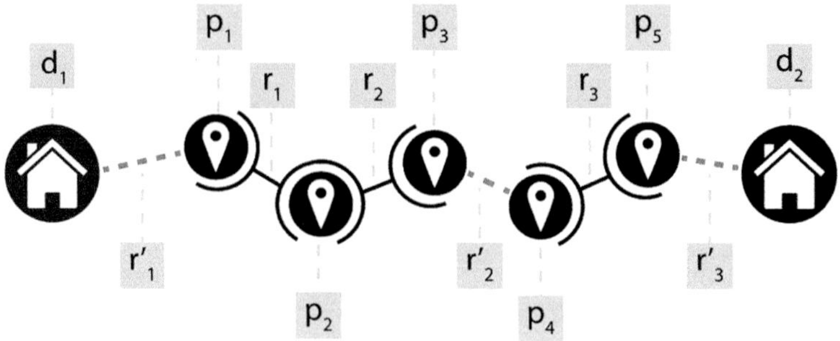

Fig. 1 Schema for a potential vehicle shift

one of $p_e(r_i)$ and $p_s(r_j)$ or the time to move from $p_e(r_i)$ to $p_s(r_j)$ is such that parking is not needed); *compound transfer* (if it is none of the cases listed above, it is necessary to perform two simple transfers: one from $p_e(r_i)$ to a parking point p_p and one from p_p to $p_s(r_j)$).

Based on the above definition, the schema for a potential vehicle shift ω_v for a given vehicle v is represented in Fig. 1, where p_1, p_2, p_3, p_4 and p_5 are stops; d_1 and d_2 are depots; r_1, r_2 and r_3 are trips; r_1', r_2' and r_3' are dead rides. More specifically r_1' is the pull-out from the depot d_1, r_3' is the pull-in to the depot d_2 and r_2' is a simple transfer made to move v from p_3 to p_4. Generally, it is not needed for a route to start and end at the same depot ($d_1 \neq d_2$). The condition to be ensured is to have the same number of vehicles of each type in every depot within the end of the planning horizon. However, in real cases the depots are not geographically distributed, therefore, imposing that $d_1 = d_2$ does not produce a significant deterioration of the solution.

2.1.1 Literature Overview

Heuristic approaches for the MDVSP problem were developed since 1970. These approaches can be grouped in heuristics based on MIP formulations and heuristics based on local search techniques. All the methods reaching good solutions within large computing times can be classified in the first group. The second group collects all the local search techniques providing quickly solutions for large size instances of the problem. More precisely, the MIP heuristics are based on column generation [19], branch and bound [4] and/or Lagrangian relaxation [20]. For an in-depth discussion on column generation for Mixed Integer Linear Programming (MILP) problems, the reader is referred to [1]. A different approach to solve the MDVSP problem is based on the use of local search [14] or tabu search techniques [8]. A heuristic approach in which the main problem is split into $|D|$ subproblems is proposed in [16]. Each of these problems is a Single Depot Vehicle Scheduling Problem that can be solved in polynomial time. The paths with the highest frequency are considered optimal and the new problem consists in allocating the remaining trips.

2.1.2 Problem Description

In practical contexts a solution of MDVSP must satisfy several constraints to be feasible. In our formulation all the following constraints are considered:

1. each trip $r \in R$ is assigned to exactly one compatible vehicle;
2. the set of trips assigned to each vehicle must be compatible with his autonomy (if the autonomy is not sufficient to complete the trips, then a refuel is required);
3. the length (expressed in km) of each activity without passengers must be smaller or equal to a given value l_{max};
4. each vehicle $v \in V$ can stand in a bus stop that is not a parking stop for a limited time;
5. each depot $d \in D$ contains a given number of vehicles for each type;
6. each vehicle $v \in V$ must perform its route within the time window of the depot in which it starts the service;
7. each vehicle $v \in V$ starts and ends its route in the same depot;
8. for each pair of consecutive trips (r_i, r_j) performed by the same vehicle, the time interval between $t_s(r_j)$ and $t_e(r_i)$ must be greater than the travel time $t^v_{i,j}$.

The mathematical formulation used to model this problem is reported in [6]. The problem is an IP problem whose feasible region is defined by $O(|R| \cdot |V|)$ constraints, some of which are non linear constraints, and by $O(|R|^2 \cdot |V|)$ binary variables. Therefore, this problem cannot be solved to optimality in case of large size instances, even in case of linearization of the non linear constraints. However, the structure of this problem can be used to compute a lower bound for the proposed model, in which constraints related to the possibility to refuel or not the vehicles are not taken into account. In fact, the relaxed model reduces to an assignment problem with the introduction of route constraints and constraints on the number of vehicles to use.

2.1.3 Solution Method

The MDVSP problem is a NP-Hard problem [17]. Therefore finding a good solution in a reasonable computational time implies using heuristic techniques. We propose an Iterated Local Search (ILS) algorithm to find a MDVSP solution, in which a neighborhood search followed by a perturbation phase are iterated as long as an exit criterion is met. In Algorithm 1, we provide a pseudocode of the overall algorithm. The various steps will be detailed in the following sections by analyzing separately the initialization, the local search, the shaking and the acceptance decisions. A starting solution is generated through the algorithm presented in the following section. Then, different local search operators are used to explore the solution space. We denote as $LocalBestSolution(type, s)$ the best solution identified by using the local search operator $type$ starting from the solution s, and we denote as $Neighborhood(type, s)$ the set of the solutions inside the neighborhood of s with respect to the local search operator $type$. The local search operators are not analyzed exhaustively but randomly within a given time interval which was tuned according to the results achieved on a subset of instances [6] where each local search operator was used separately.

Algorithm 1 Iterated Local Search Algorithm for the MDVSP

Require: Set of trips R, vehicles V, depots D, bus stops P
Calculation of the lower bound value f_{lb}
Generation of a starting solution s with value $f(s)$
Set $IT_{current}$, $t_{current} = 0$, $f^* = f(s)$, $s_{current} = s$ and $s^* = s$
while $IT_{current} < IT_{max}$ and $t_{current} < t_{max}$ and $f^*/f_{lb} > \varepsilon$ **do**
  Set $f(s_l) = +\infty$
  for $type \in localSearchTypes$ **do**
    $s_t = LocalBestSolution(type, s)$
    if $f(s_t) < f(s_l)$ **then**
      $f(s_l) = f(s_t)$
    end if
  end for
  if $f(s_l) < f(s_{current})$ **then**
    $s_{current} = s_l$
  end if
  if $f(s_l) < f^*$ **then**
    $s^* = s_l$
    $f^* = f(s_l)$
  else
    for $s_l \in Neighborhood(Nshift, s)$ **do**
      Set $p' = N(0, 1)$
      Set $p = e^{\left(\frac{f(s_l)-f^*}{T}\right)}$
      if $p' < p$ **then**
        $s_{current} = s_l$
      end if
    end for
  end if
  Update T, $t_{current}$, $IT_{current}$
end while

2.1.4 Initial Solution

In this phase each trips is iteratively assigned to one vehicle at the least cost. Before starting the algorithm, the trips are sorted according to their departure times, if two trips start at the same instant the one with the lowest arrival time is examined first. The rationale of using this criterion is motivated by the fact that the feasibility of assigning a trip to a vehicle is performed before by checking its workload. At each iteration of the algorithm only one trip is examined. The first step consists of identifying the list of vehicles that can be assigned to the current ride r_i. The vehicle v can perform the ride r_i if the following conditions are satisfied: (a) it belongs to a compatible type, (b) it has sufficient time and fuel to perform the trip, (c) constraints on the maximum length of an activity without passengers are not violated. Under these conditions, trip r is assigned to a suitable vehicle v' that minimizes the increment of the cost function $\Delta fo : v' = argmin_{v \in V} \Delta fo_v$, where Δfo_v is calculated as follows:

$$\Delta fo_v = \begin{cases} l^v_{i,j}, & \text{if } |\omega_v| \geq 0 \\ l^v_{d,j} + K, & \text{otherwise} \end{cases} \tag{1}$$

2.1.5 Local Search

The solution found by the algorithm above described is generally not optimal. Hence, different local search operators are used with the aim of improving the initial solution. A feasibility check is performed whenever a new solution is analyzed in order to ensure that the services of the involved vehicles remain feasible. The following local search operators were used to improve the initial solution:

- *Nshift*: starting from a solution s, this operator builds a neighborhood by considering all the solutions that are obtained by shifting a trip between two vehicles.
- *Nswap*: this operator moves trip r_j serviced by vehicle v_i to vehicle v_h, and trip r_k serviced by vehicle v_h to vehicle v_i. The aim is to find the best reduction of the objective function value.
- *Blocks*: This operator builds a neighborhood by splitting a sequence B of consecutive trips $B = \{r_j, r_k, \ldots, r_z\}$ serviced by vehicle v_i, and assigning them to one or more different vehicles. The size of each block is generated randomly at each iteration within a range that depends on the specific instance.
- *2opt** In the $2opt^*$ technique the services performed by vehicles v_i and v_h are split in two parts. The solution will be modified if it is possible to obtain a reduction of the objective function by joining the first part of the service performed by v_i to the second part of the service performed by v_h, and vice versa.
- *Changing vehicles*: in the changing vehicle procedure all trips performed by vehicle v_i are assigned to vehicle v_h. If vehicle v_h already contains some trips, the existing trips will be merged with the news ones to reconstruct the route traveled by v_h.

2.1.6 Perturbation and Stop Criteria

We use the simulated annealing paradigm to escape from local optima. Simulated annealing is a probabilistic meta-heuristic used to find a good approximation of the global optimum of a given function [15]. At each iteration, an exploration of the feasible region around the current solution s, denoted by $N(s)$, is performed through the operators described in the previous sections. If the best solution $s' \in N(s)$ has a cost lower than the cost of s, this solution is accepted and the current optimal value is updated. However, even if this condition is not met the solution s' is accepted under the following condition:

$$e^{-\left(\frac{f-f'}{T}\right)} \leq p' \tag{2}$$

where f and f' are respectively the cost of the current and the new solution, respectively. The value p' is a random number with a uniform distribution between [0, 1], whereas T is the so-called temperature, whose value decreases during the iterations. It is crucial to find the starting value of T because high values allow accepting a large number of bad solutions. On the other hand, small values limit the capability of the algorithm to explore different portions of the solution space. An appropriate starting value of parameter T is chosen such that the initial probability p ranges between 0.6 and 0.9. According to our extensive preliminary testing, we found it profitable to use different criteria to stop the ILS algorithm. The following criteria are used: the processing time exceeds a maximum time; the number of performed iterations exceeds a maximum number; the gap between the best objective value and the lower bound is smaller than a given threshold. The temperature value decrease in a linear way during the execution of the algorithm according to the residual time and iterations reaching a value equal to 0 at the end.

2.2 *Crew Scheduling Problem*

The salary of the drivers usually covers a percentage between 50 and 75% of the total costs. Therefore it is important to minimize the number of drivers using efficiently all the available resources. The crew scheduling problem presents as many similarities with the MDVSP as many differences due to the presence of constraints modeling safety conditions and law requirements. The trips assigned to each driver are grouped as follows:

1. driver-trips: all the trips assigned to a given vehicle driver.
2. passenger-trips: all the trips assigned to a vehicle that carries the driver as passenger. Each passenger-trip must be concatenated with a driver-trip assigned to the same or a different vehicle driven by this driver. The passenger-trips are associated with activities in which a driver must be moved in a geographical area to execute trips.

Each driver belongs to a given *Service Unit*. A Service unit U is composed by a set of bus stops where the driver can get on or get off the vehicle. The shift ω_c of a driver c is defined by n activities chronologically sorted. A shift can be divided into three subsets, named D, B, P that represent the set of driver-trips, breaks, and passenger-trips, respectively. Let c_N denote the time between the beginning of the first activity and the end of the last. Let c_w denote the working time defined as the time period in which the driver is paid by the company, c_w is equal to c_N minus all the time breaks included in B longer than μ, where μ is a given time. The following definitions will be helpful in the rest of the paper:

1. maximum driving time c_D;
2. maximum driving time without a break $c_{\overline{D}}$: the time between two consecutive activities is considered a break only if it is greater than μ. Based on the value of μ, set B is partitioned as follows: $B_{\geq\mu} = \{a \in B : t(a) \geq \mu\}$ and $B_{<\mu} = \{a \in B : t(a) < \mu\}$. The driving time without a break is calculated (as reported in the EU regulation no. 561/2006).

2.2.1 Literaure Overview

The urban bus crew scheduling problem has received a great attention in the last decade, and several model and solving methods were developed. The most used models can be classified into set covering, set partitioning [18] and multi-objective models [12]. These models were solved either through exact methods like column generation and branch and bound or heuristic algorithms like tabu search, genetic algorithm and colony algorithm. An algorithm based on a set partitioning model is proposed in [21] while a greedy approach based on genetic algorithms is presented in [7]. In the last decade many studies based on an integrated approaches for the vehicle and crew scheduling have been presented [10, 11, 13]. In this context, our approach is different from the ones presented in the literature. Such a difference relies on the fact that the integration between the MDVSP and the CSP is designed to consider that finding a solution for the second problem implies to consider more realistic constraints that modify the solution found for the first problem.

2.2.2 Probem Description

Constraints that a bus driver shift must comply with are specified as follows:

- each driver starts and ends the shift in a bus stop that belongs to the corresponding Service Unit;
- each driver starts the service within time window $S_c = [c_m^S, c_M^S]$ and ends the service in time window $E_c = [c_m^E, c_M^E]$;
- the distance traveled by a bus driver is limited;
- a minimum number of breaks, named b_{min}, must be assigned to each bus driver;
- a bus driver can change vehicle only on a subset of bus stops, named relief points.

The mathematical formulation used to model this problem is reported in [6]. The model contains integer variables and has a size (number of variables and constraints) excessively large to be solved exactly. Therefore, finding a good solution of this problem can be performed by using heuristic approaches. However also in this case, the structure of this problem can be used to compute a lower bound of the crew scheduling problem, in which some constraints are not considered. More specifically, in the relaxed model we relaxed the constraints related to the service unit and the minimum/maximum driving time.

2.2.3 Solution Method

The outline of the algorithm used to find a solution of the CSP problem is similar to the one proposed for the MDVSP. Initially, a feasible solution is found using a greedy approach and then such a solution is improved using local search operators. However the main difference in this algorithm is the link between the two problems. The local search techniques is devoted to find a solution of the MDVSP problem that minimizes either the number of vehicles or the number of drivers.

2.2.4 Initial Solution

The basic idea in the algorithm is to generate a number of tasks and assign a subset of these tasks to the drivers in such a way that each trip is assigned to at least one driver. A task T is a set of consecutive trips performed by the same vehicle that starts and ends in a relief point that belongs to the same Service Unit. To ensure the respect of the constraints on the maximum driving time without a break, only tasks with a length smaller of $c_{\overline{D}}$ are considered. At each iteration a task and a driver are selected by minimizing the following quantity:

$$\frac{w_1 \cdot \Delta c_N + w_2 \cdot \Delta c_W + w_3 \cdot \Delta c_D}{|R_T|} + \Delta f_c \tag{3}$$

where $|R_T|$ is the number of not yet assigned trip in the task, w_1, w_2 and w_3 are weights assigned to c_N, c_W and c_D, whereas f_c is a fixed cost that is paid if c_N is greater than 0. Trips already assigned are not considered in the denominator of this function and they are considered like passenger trips for the driver. The algorithm ends when all the trips are assigned to at least one driver or is not possible to assign a new task. The solution returned by this procedure could be far from the optimum, but also infeasible. Therefore, to restore the feasibility and improve the quality of the solution several well-known local search techniques and a new operator, named $Shiftunion$, are used.

2.2.5 Local Search

The initial solution could use a large number of drivers. A reduction of this number represents a goal for the company. The $ShiftUnion$ operator aims at joining the shifts of two drivers without violating constraints More precisely, given two different shifts c_1, c_2, a new shift is created when a task T exists such that $c_1 \cup c_2 \cup T$ is feasible. The task T is used to guarantee that each activity starts in the same position in which the previous activity ended. The shift returned by the union of two drivers may have a length c_N equal, lesser (by reducing the total time in which the driver does not perform any ride) or greater than the sum of the joined shifts. In some cases it is possible to join the shift of two drivers with no new activity. To take into account this option, a dummy task that does not contains any trips is introduced.

2.2.6 Post-optimization Algorithm

After generating a crew scheduling solution, an improvement is executed by using the local search techniques illustrated in the previous sections. The output data of the MDVSP is a set of trips $R \cup R'$ and routes $\sigma = \{\sigma_{v_1}, \sigma_{v_2}, \ldots, \sigma_{v_m}\}$. Set R does not depend on the solution, while R' changes by altering the allocation of the trips in the MDVSP. A solution of the MDVSP is feasible if each trip is assigned to exactly one vehicle. The output data of the CSP is a set of shifts $\{c_1, c_2, \ldots, c_n\}$ that is assigned to each driver $c \in C$, such that a trip can be assigned to more than one driver. The local search techniques are performed by considering also the new allocation of the trips to the drivers. The operators $ShiftAllocation$ and $ShiftUnion$ (with which each new generated dead runs is assigned to the set of drivers at the minimum cost) are executed on each current solution, and the current optimum is updated whenever the sum of the objective functions of the MDVSP and the CSP is reduced. The feasibility of the overall solution is ensured by the procedure that is outlined in Algorithm 2.

Algorithm 2 Global optimization

Require: Set of trips $R \cup R'$, vehicles V, drivers C and a feasible solution s_0

$s = s_0 \rightarrow$ current solution
$z_0 \rightarrow$ objective function of the current solution
$r_j \rightarrow$ select a random trip $\in R$
$v_i \rightarrow$ current vehicle of r_j
put $R_{v_i}=R_{v_i} - r_j$
if R_{v_i} is feasible **then**
  for $v \in V$ **do**
    put $R_v=R_v \cup r_j$
    $R'_{new} \rightarrow$ new set of activities without passenger
    for $r \in \{R' - R'_{new}\}$ **do**
      $c \rightarrow$ current driver of r
      put $R_c=R_c - r$
    end for
    for $r \in \{R'_{new} - R'\}$ **do**
      ShiftAllocation(r,C)
    end for
    ShiftUnion(C)
    if s is not feasible **or** $z(s) > z_0$ **then**
      $s = s_0$
    end if
  end for
else
  put $R_{v_i}=R_{v_i} \cup r_j$
end if

3 Computational Experiments

This section illustrates the computational results obtained running the algorithm on a data set of instances related to a public transport company. The goal is to prove the effectiveness of the proposed ILS (Iterated Local Search) algorithm. The main computational analysis is clearly devoted to the logistic problem in which a MDVSP and a CSP are integrated together. As far as we know, no benchmark instances exist in the literature for an integrated problem like the one under analysis. Therefore, we generated three sets of instances, different for numbers of trips, depots and bus and park stops:

- small-sized (Id problem 1–15) instances, with not more than 150 trips drivers are solved by using the ILS algorithm. Solutions are compared with the ones of lower bound models solved with a general purpose optimization software;
- medium-sized instances (Id problem 16–30) with up 300 trips, These instances were designed from real data provided by Italian transport companies;
- large-sized instances (Id problem 31–45), with more than 300 trips. These instances were designed by taking into account as many data as possible coming from the available real data.

The ILS algorithm was implemented as single thread code in Java and all tests are performed on a desktop computer equipped with an Intel Core i3 processor with 2.2 GHz, 6 GB RAM, and running Windows 7 Professional. For each instance we impose the following limits for each vehicle service: autonomy of the vehicle equal to 500 km; maximum length of an activity without passenger equal to 40 km; average speed of the vehicle for activities without passenger equal to 40 km/h; maximum time that a vehicle can stand in a bus stop where parking is not allowed equal to 3 min; a vehicle can perform a refuel at any depot. In these computational tests the results of the MDVSP problems are used as input data for the CSP problems, where each trip has to be assigned to a driver. For each driver, the following parameters are given: maximum distance walked by a bus driver equal to 0.5 km; maximum length of the shift equal to 12 h; maximum working time equal to 8 h; maximum driving time equal to 6 h; maximum driving time without a break equal to 2.5 h; minimum length of a break equal to 20 min; minimum number of breaks assigned to a driver equal to 2; maximum number of breaks assigned to a driver equal to 4. Moreover, in the following tests, three classes of shifts are considered. Different time windows are used:

- morning shift: each driver of this class begins its turn within the time window [0:00–10:00] and ends to work within the time window [8:00–18:00];
- afternoon shift: each driver of this class begins its turn within the time window [8:00–18:00] and ends to work within the time window [16:00–24:00];
- evening shift: each driver of this class begins its turn within the time window [16:00–24:00].

The results of the proposed ILS algorithm for the MDVSP and the CSP are compared with the lower bounds obtained using the model relaxations. For all the instances

Table 1 Summary table of the results achieved

Id problem	Vehicles	km transfer	Drivers	Working time	z	$Gap\%$	$\mu_{Gap\%}$
1	11	295.40	12	89.97	2945.15	2.68	7.36
2	16	411.38	16	123.81	4049.44	3.28	7.45
3	14	403.72	15	102.95	3633.22	4.13	8.97
4	19	829.31	26	172.99	6109.22	9.07	12.12
5	22	649.29	25	170.52	5954.52	7.20	10.80
6	20	423.53	22	162.75	5251.05	7.57	11.56
7	22	588.20	24	184.62	5934.40	13.03	15.64
8	21	558.59	22	161.27	5421.30	3.61	8.57
9	22	449.20	26	196.58	6115.03	4.85	9.73
10	21	182.78	24	185.23	5485.05	2.23	5.12
11	23	171.41	24	173.45	5455.91	3.13	7.90
12	23	604.14	24	159.49	5749.06	4.90	7.76
13	21	367.49	22	158.44	5201.86	9.02	12.15
14	22	619.04	19	149.82	5117.29	9.68	12.82
15	21	505.95	18	129.06	4646.57	3.92	8.51
16	14	858.20	16	118.97	4347.94	4.32	8.41
17	21	1373.72	23	182.32	6546.91	2.94	5.77
18	38	1806.05	44	326.43	11370.37	15.65	17.78
19	31	1192.13	36	275.87	9100.85	13.22	17.50
20	37	1734.24	44	299.58	10980.07	9.37	12.53
21	36	833.29	43	283.46	9767.94	3.74	6.22
22	32	1367.28	43	289.41	10161.40	12.75	15.20
23	27	995.50	39	296.90	9214.49	10.18	15.12
24	26	1411.18	33	225.55	8266.66	7.10	11.07
25	33	1190.99	38	279.02	9431.18	10.72	14.01
26	29	1451.93	33	248.76	8689.50	6.96	10.79
27	29	1472.32	32	248.35	8605.84	9.80	11.84
28	28	1472.02	34	233.13	8603.37	8.27	11.52
29	31	1217.91	38	256.18	9129.68	3.40	6.73
30	32	983.05	38	264.90	9032.03	9.75	12.13
31	35	303.83	58	392.64	11780.26	6.01	10.93
32	22	1284.78	48	352.22	10706.97	8.95	12.34
33	21	789.66	47	313.88	9678.51	4.70	7.20
34	43	2555.55	59	420.35	14809.06	7.18	12.09
35	46	1678.58	64	467.53	15053.91	9.04	13.27
36	43	428.71	62	409.42	12872.88	7.60	10.52
37	42	879.16	57	409.43	12773.47	2.96	6.05

(continued)

Table 1 (continued)

Id problem	Vehicles	km transfer	Drivers	Working time	z	$Gap\%$	$\mu_{Gap\%}$
38	45	843.36	68	472.85	14621.88	9.12	12.41
39	41	1236.40	63	459.56	14182.00	4.31	8.44
40	45	901.14	66	435.16	14102.78	7.15	9.58
41	43	833.73	59	394.71	12830.83	2.01	5.46
42	39	744.49	63	444.69	13441.41	2.88	7.53
43	41	1163.61	65	481.34	14527.06	5.89	8.98
44	40	858.58	58	404.72	12705.82	2.28	7.18
45	40	932.35	56	446.09	12993.25	4.34	6.91

we used a fixed cost K equal to 50 for each vehicle, a fixed cost equal to 100 for each driver and a cost equal to 10 for each additional working hour. The lower bound z_{lb} is used to calculate the percentage gap, denoted by $Gap\%$, with respect to the objective function of the best solution returned by the algorithm, named z_{ILS}. This gap is computed as follow: $Gap\% = \frac{z_{ILS}-z_{lb}}{z_{lb}}$, where z_{lb}^{MDVSP} denotes the lower bound for the multi depot vehicle scheduling problem, z_{lb}^{CSP} denotes the lower bound of the crew scheduling problem, and z_{lb} indicates the overall lower bound calculated as the sum of the bounds for both problems. Due to the stochastic nature of the algorithm, 10 runs are performed for each instance. A time limit of five minutes is imposed to every execution of the algorithm related to the MDVSP, and an additional time of five minutes is given to optimize the overall solution. The results are reported in Table 1 where the columns indicate the number of vehicles, the number of kilometers for transfer activities, the number of drivers, the $Gap\%$ of the best solution and the average percentage gap, denoted by ($\mu_{Gap\%}$) obtained in 10 runs.

The table shows that the percentage gap of the best solution ranges between 2.01 and 15.65%. However, it is possible to see that the largest gaps are reached for the instances with 26 depots. The largest gaps are partially due to the deterioration of the lower bound for the MDVSP problem in the instances where the number of depots is high, because the length of the pull-in and pull-out trips are always evaluated by considering the nearest depot.

4 Conclusions

The resolution of *Multi Depot Vehicle Scheduling Problems* and *Crew Scheduling Problems* is an extremely important step for the operation planning of a local transport company. Since 1970 different heuristic algorithms have been proposed for both problems. However only in the last decade the two problems have been solved with an integrated approach. Moreover these approaches solve problems based on several simplifying assumptions. The aim of this paper is to present an integrated algorithm for the MDVSP and the CSP problems where different constraints according to EU legal framework (561/2006) have been considered. This technique is based on the principle that a sequential approach can be used to reduce the computational time using some procedures that allow to consider the connections of these two problems. The tests made show that a significant improvement of a starting solution can be obtained using tailored local search techniques. Moreover, Simulated Annealing was used to explore different zone of the solution space and escape from local minima. The results demonstrate the quality of the proposed method, that can be used to determine in a small time a solution for real context with a significative number of trips, depots, vehicles and drivers.

References

1. Barnhart, C., Johnson, E.L., Nemhauser, G.L., Savelsbergh, M.W., Vance, P.H.: Branch-and-price: column generation for solving huge integer programs. Oper. Res. **46**(3), 316–329 (1998)
2. Beasley, J.E., Cao, B.: A tree search algorithm for the crew scheduling problem. Eur. J. Oper. Res. **94**(3), 517–526 (1996)
3. Branke, J., Schmidt, C., Withopf, M.: A fast look-ahead heuristic for the multi-depot vehicle routing problem. In: Wirtschaftsinformatik (2), pp. 411–428. Citeseer (2007)
4. Carpaneto, G., Dell'Amico, M., Fischetti, M., Toth, P.: A branch and bound algorithm for the multiple depot vehicle scheduling problem. Networks **19**(5), 531–548 (1989)
5. Carraresi, P., Gallo, G.: Network models for vehicle and crew scheduling. Eur. J. Oper. Res. **16**(2), 139–151 (1984)
6. Ciancio, C., Laganà, D., Musmanno, R., Santoro, F.: An integrated algorithm for shift scheduling problems for local public transport companies. Omega **75**, 139–153 (2018)
7. Clement, R., Wren, A.: Greedy genetic algorithms, optimizing mutations and bus driver scheduling. In: Computer-Aided Transit Scheduling, pp. 213–235. Springer (1995)
8. Cordeau, J.F., Gendreau, M., Laporte, G.: A tabu search heuristic for periodic and multi-depot vehicle routing problems. Networks **30**(2), 105–119 (1997)
9. Crevier, B., Cordeau, J.F., Laporte, G.: The multi-depot vehicle routing problem with inter-depot routes. Eur. J. Oper. Res. **176**(2), 756–773 (2007)
10. Fores, S., Proll, L., Wren, A.: Experiences with a Flexible Driver Scheduler. Springer (2001)
11. Freling, R., Huisman, D., Wagelmans, A.P.: Models and algorithms for integration of vehicle and crew scheduling. J. Sched. **6**(1), 63–85 (2003)
12. Ghoseiri, K., Szidarovszky, F., Asgharpour, M.J.: A multi-objective train scheduling model and solution. Transp. Res. Part B: Methodol. **38**(10), 927–952 (2004)
13. Huisman, D., Freling, R., Wagelmans, A.P.: Multiple-depot integrated vehicle and crew scheduling. Transp. Sci. **39**(4), 491–502 (2005)
14. Jin-Kao Hao, B.L.: Iterated local search for the multiple depot vehicle scheduling problem. Comput. Ind. Eng. **57**(1), 277–286 (2009)
15. Kirkpatrick, S.: Optimization by simulated annealing: quantitative studies. J. Stat. Phys. **34**(5–6), 975–986 (1984)
16. Kliewer, N., Mellouli, T., Suhl, L.: A time-space network based exact optimization model for multi-depot bus scheduling. Eur. J. Oper. Res. **175**(3), 1616–1627 (2006)
17. Lenstra, J.K., Kan, A.: Complexity of vehicle routing and scheduling problems. Networks **11**(2), 221–227 (1981)
18. Marsten, R.E., Shepardson, F.: Exact solution of crew scheduling problems using the set partitioning model: recent successful applications. Networks **11**(2), 165–177 (1981)

19. Oukil, A., Amor, H.B., Desrosiers, J., El Gueddari, H.: Stabilized column generation for highly degenerate multiple-depot vehicle scheduling problems. Comput. Oper. Res. **34**(3), 817–834 (2007)
20. Pepin, A.S., Desaulniers, G., Hertz, A., Huisman, D.: Comparison of heuristic approaches for the multiple depot vehicle scheduling problem. Technical Report, Erasmus School of Economics (ESE) (2006)
21. Yunes, T.H., Moura, A.V., De Souza, C.C.: Solving very large crew scheduling problems to optimality. In: Proceedings of the 2000 ACM Symposium on Applied Computing, vol. 1, pp. 446–451. ACM (2000)

A Tool for Practical Integrated Time-Table Design and Vehicle Scheduling in Public Transport Systems

Samuela Carosi, Antonio Frangioni, Laura Galli, Leopoldo Girardi and Giuliano Vallese

Abstract Planning of services and resources for public transport systems is a complex process. In practice the planning is usually decomposed into several stages, where service levels are decided first, while vehicles and drivers are only optimized later on. We describe the new `TTD.XT` tool, developed by M.A.I.O.R. S.r.l. in collaboration with the University of Pisa, that improves the efficient utilization of expensive resources by *simultaneously* optimizing timetabling and vehicle scheduling. We quickly present the underlying mathematical model and (math-heuristic) algorithm, and compare the solutions constructed by the tool for real-world bus transport planning instances for three major Italian cities with those produced by the standard sequential decision process; the former are significantly better than the latter both in terms of objective function value and when judged by experts of the field.

Keywords Timetabling · Vehicle-scheduling · Integrated approach · Matheuristic

1 Introduction

Public transportation companies often face complex logistic problems. Following a "divide et impera" approach, it is customary to subdivide such large-scale problems into several *stages*, that are solved in sequence. Stages are not independent, but

S. Carosi · L. Girardi · G. Vallese
Department of OR Development, M.A.I.O.R. S.r.l., Lucca, Italy
e-mail: samuela.carosi@maior.it

L. Girardi
e-mail: leopoldo.girardi@maior.it

G. Vallese
e-mail: giuliano.vallese@maior.it

A. Frangioni (✉) · L. Galli
Dipartimento di Informatica, Università di Pisa, Largo B. Pontecorvo 3, 56127 Pisa, Italy
e-mail: antonio.frangioni@unipi.it

L. Galli
e-mail: laura.galli@unipi.it

M. Dell'Amico et al. (eds.), *A View of Operations Research Applications in Italy, 2018*, AIRO Springer Series 2, https://doi.org/10.1007/978-3-030-25842-9_16

strictly interconnected, so that the solution of one stage affects all the subsequent ones; therefore, planning in sequence is a *heuristic* strategy, that generally produces suboptimal solutions. In urban public transport, the usual stages are: *network design*, *timetabling* (TT), *vehicle scheduling* (VS), and *crew scheduling*. The first two define the "quality" of the service to be offered, while the last two are devoted to resource allocation. Thus, expensive resources for the companies, such as the vehicles and drivers needed, are only determined in the later stages of the planning process. We describe the new `TTD.XT` tool developed by M.A.I.O.R. S.r.l., a leading company providing services and advanced software systems to public transport authorities and operators, in collaboration with the Department of Computer Science of the University of Pisa. The tool is based on a new model and algorithm for the Integrated Timetabling and Vehicle Scheduling (ITTVS) problem, that jointly optimizes a service stage (TT) and a resource allocation stage (VS). The approach was tested on several real-world instances originating from M.A.I.O.R. customers, and proved to be capable of producing good-quality solutions—if compared with those manually constructed by experts of the transport companies and actually used in operations—with running-times compatible with the user requirements.

The structure of this chapter is as follows. Section 2 presents the application, with a brief discussion of the existing literature to underline the specific contribution of our research. A "base case" of the problem is mathematically formulated in Sect. 3, and a *matheuristic* approach is described; the tool developed in M.A.I.O.R. actually extends the model with a number of other features required by the customers that cannot be described here for space reasons. Computational results are discussed in Sect. 4 which show the usefulness that the tool may have for planners.

2 Problem Description

For space reasons we only describe the "base case" ITTVS problem at M.A.I.O.R., consisting of a *public transportation network* (PTN) with a *simple single line* topology; both the mathematical formulation and the algorithmic approach extends to more complex PTN topologies with multiple independent lines, as well as to other features (tighter control on the vehicles entering/leaving the depot, introduction in the vehicle schedule of features useful to construct a feasible driver schedule, ...) that M.A.I.O.R. customers require, as described in [3]. The main input to ITTVS is a graph, where the nodes correspond to bus stops or depots, and the links correspond to direct bus transits, a set of potential trips $\mathscr{T}$, the desired headways, the bus travel times and the stopping times at the depot/terminals. A *simple (single) line* PTN topology is just a path AB between two *terminals* A and B (i.e., start/end stops of the line), which typically has the *in-bound* and *out-bound directions* $\mathscr{D} = \{\mathbf{AB}, \mathbf{BA}\}$ (but circular lines with only one direction exist). In more general cases, a single line may comprise multiple routes for each direction. A *trip* $i \in \mathscr{T}$ corresponds to a direction $d(i)$, has start and end terminal $sn(i)$ and $en(i)$, respectively, with the corresponding departure time $st(i)$ and arrival time $et(i)$, respectively. Each trip can (but need not necessarily) be operated by some vehicle; since each trip belongs to a given direction, we define $\mathscr{T} = [\mathscr{T}_d]_{d\in\mathscr{D}}$ as the "direction partition" of $\mathscr{T}$. In order to model the VS sub-problem, it is also necessary to consider the *single deposit* node O in the PTN, but not any other intermediate stop of the line. In the following, we will denote by N the set of terminals of the involved lines (say, $N = \{A, B\}$ in the simple case) and by $N^+ = N \cup \{O\}$. For each direction $d \in \mathscr{D}$, a *main stop* is identified, which is used to calculate the *headways*, as detailed below.

The Timetabling Sub-problem

Generally stated, a TimeTable (TT) is a subset of trips that meets some service frequency requirements. To define *regularity* of a TT, the relevant information is the arrival time $a(i)$ at the main stop of $d(i)$ (the same for all the vehicles, since we assume only one vehicle type), which is used to calculate the *headways*. A *TT* π_d *for a direction* $d \in \mathscr{D}$ is a subset of its potential input trips $\mathscr{T}_d$ that satisfies headway constraints; since the latter are defined "per direction", the TT sub-problem can be decomposed accordingly. Thus, a TT is just the union of $|\mathscr{D}|$ (independent) TT, one for each direction, i.e., $\pi = [\pi_d]_{d\in\mathscr{D}}$. Given a TT π, the (actual) headways of a direction d w.r.t. π are the times separating each two consecutive trips i, j in π_d passing at its main stop, i.e., $a(j) - a(i)$. Timetabling is non-periodic, i.e., we are given a time horizon T (say 5:00–24:00); as the *ideal headway* varies along the day, T is partitioned into *time windows*. We will denote by $h(i)$ the time window in which trip i happens, assuming for simplicity that trips are completely contained in a time window (only minor changes are required to account for "border effects"). For each time window h and each direction $d \in \mathscr{D}$, we are given the *minimum, ideal and maximum headways* $\underline{I}_d^h \leq I_d^h \leq \bar{I}_d^h$; thus, $\pi_d \subset \mathscr{T}_d$ is a *feasible TT* if $a(j) - a(i) \in [\underline{I}_d^{h(i)}, \bar{I}_d^{h(i)}]$, i.e., the headway of each two consecutive trips i and j in π_d is feasible. Also, the *first* and the *last trip* of π_d belong to given subsets $\mathscr{T}_d^{ini}$ and $\mathscr{T}_d^{fin}$ of *initial* and *final* trips, specified as an input to the problem. The quality of a TT is evaluated via a *penalty function* of the deviation of a feasible actual headway $\bar{a} \in [\underline{I}_d^h, \bar{I}_d^h]$ from the ideal one I_d^h. The actual form of this function is immaterial for our model, just assuming the trivial properties that the penalty is zero if $\bar{a} = I_d^h$, and larger than zero (typically, nondecreasing in $|\bar{a} - I_d^h|$) otherwise.

The Vehicle Scheduling Sub-problem

A feasible schedule for a single vehicle is a subset of *compatible* trips, i.e., that can be performed by the same vehicle while satisfying the required temporal and legal restrictions. A vehicle leaving a depot to reach the start-terminal of a trip is said to be performing a *pull-out trip*; similarly, it performs a *pull-in trip* when it returns to the depot from the end-terminal of a trip. For each terminal $n \in N$ and for each time window h, we are given the *travel time* for a pull-in and pull-out trip, denoted by t_{n+}^h and t_{n-}^h, respectively. Note that pull-in and pull-out trips are not included in $\mathscr{T}$, as they are not passenger service trips. For each node $n \in N^+$ and for each time window h we are given a *minimum* and a *maximum stopping-time*, denoted by $\delta_{min,n}^h$

and $\delta^h_{max,n}$, respectively; however, we typically assume $\delta^h_{max,O} = +\infty$ for all h. Trips $i, j \in \mathcal{T}$ are *compatible* if they can be covered consecutively by the same vehicle:

- trip j starts at the terminal where i ends ($en(i) = sn(j)$), and the waiting time at the terminal between the end of trip i and the start of trip j is feasible ($\delta^{h(i)}_{min,en(i)} \leq st(j) - et(i) \leq \delta^{h(i)}_{max,en(i)}$), in which case they are said *in-line compatible*;
- there is enough time between the end of trip i and the start of trip j to perform a pull-in trip from $en(i)$, wait the minimum amount of time at the depot, and then perform a pull-out trip towards $sn(j)$ ($en(i) \neq sn(j)$ and $st(j) - et(i) \geq t^{h(i)}_{en(i)+} + \delta^{h(i)}_{min,O} + t^{h(j)}_{sn(j)-}$), in which case they are said *out-line compatible*.

Thus, a feasible schedule for a vehicle has the following structure: an initial pull-out trip to leave the depot, a sequence of compatible (service) trips in $\mathcal{T}$, possibly separated by pull-in/out trips, and a final pull-in trip to return to the depot. A feasible *Vehicle Schedule* (VS) Ω is a subset of the input potential trips $\mathcal{T}$ partitioned into feasible schedules for single vehicles. The primary objective to evaluate the quality of a VS is the number of vehicles used. However, a secondary objective that is also important in practice is the time spent by the vehicles waiting at the terminals/depot and performing pull-in and pull-out trips.

Innovative Aspects

Almost all the articles in the literature of ITTVS focus on *transfer coordination* between multiple-lines PTN topologies, where one is interested in finding schedules that minimize some synchronization measures (transfer and/or waiting time of passengers at the stops connecting different lines). The only exception is [8], where *regularity* (i.e., minimizing the deviation from the desired headways) is considered. Also, almost all the contributions use meta-heuristics, save for [4] where a matheuristic approach is developed for an integrated bi-objective formulation, but with the transfer coordination objective function. The ITTVS model at M.A.I.O.R. is non-periodic, based on *weighting* of the bi-objective function: regularity (*service quality*) for the TT component, and *operator cost* for the single depot, single vehicle type for the VS component. To our knowledge, this is the first *matheuristic* for ITTVS with regularity objective function.

3 Mathematical Formulation and Solution Method

We present a *Mixed Integer Linear Programming (MILP)* multicommodity flow-type model for the "base case" of ITTVS; a more general version is detailed in [3]. The model consists of $|\mathcal{D}|$ *TT sub-problems*, one for each direction, a single *VS sub-problem*, and some *linking* constraints to integrate TT and VS. The TT sub-problems generate feasible timetables $\pi^* = [\pi^*_d]_{d \in \mathcal{D}}$ for all directions, that minimize the total cost of deviation from the ideal headways. The VS sub-problem constructs a feasible vehicle schedule Ω^* with minimum operator cost. Clearly, TT and VS are not independent, because the linking constraints ensure that the trips covered by Ω^*

correspond to all and only the trips used in the timetable π^*; in other words, Ω^* is a *vehicle schedule cover* of π^*.

TT Model

The TT model represents feasible subset of trips, i.e., compatible w.r.t. headway constraints. Since TT-compatibility involves pairs of subsequent trips, a timetable can be modeled in terms of *paths* on a directed *compatibility TT graph* $G_d^{TT} = (N_d^{TT}, A_d^{TT})$, where the nodes N_d^{TT} represent the trips in $\mathcal{T}_d$ plus a dummy source/sink O_d^-/O_d^+: $N_d^{TT} = \mathcal{T}_d \cup \{O_d^-, O_d^+\}$. The arcs in A_d^{TT} starting from O_d^- connect the source node to the nodes corresponding to the set of potential initial trips $\mathcal{T}_d^{ini}$, and symmetrically for the sink node O_d^+. An arc $(i, j) \in A_d^{TT}$ between two trips $i, j \in \mathcal{T}_d$ exists if and only if i and j are TT-compatible, i.e., neither "too close" or "too far apart", and its cost is computed off-line according to the given penalty function. G_d^{TT} is clearly acyclic and a path between O_d^- and O_d^+ in G_d^{TT} corresponds to a feasible timetable π_d, whose cost is the total cost of violation w.r.t. the given ideal headways.

VS Model

The VS model is based on representing feasible subsets of trips, i.e., in-line/out-line compatible. The VS-compatibility involves pairs of subsequent trips, so feasible vehicle schedules can be modeled in terms of flows on a *single* directed compatibility *VS graph* $G^{VS} = (N^{VS}, A^{VS})$, where N^{VS} contains *two* nodes i^- and i^+ for each trip $i \in \mathcal{T}$, representing the start and the end of trip i, respectively, plus two nodes O^- and O^+ representing (leaving and returning to) the deopt. As for the arc set A^{VS}, it contains six types of arcs:

1. *trip arcs* (i^-, i^+) for each trip $i \in \mathcal{T}$, with 0 cost;
2. *in-line compatibility arcs* (i^+, j^-) for each pair of trips i and j that are in-line compatible, with cost proportional to the *waiting time* at the terminals;
3. *out-line compatibility arcs* (i^+, j^-) for each pair of trips i, j that are out-line compatible, with cost proportional to the sum of the *pull-in/out travel times*;
4. *start arcs* (O^-, i^-) for each trip $i \in \mathcal{T}$, with cost proportional to the pull-out time $t_{sn(i)-}^{h(i)}$, indicating that a vehicle will perform a pull-out trip to reach $sn(i)$;
5. *end arcs* (i^+, O^+) for each trip $i \in \mathcal{T}$, with cost proportional to the pull-in time $t_{en(i)+}^{h(i)}$, indicating that a vehicle will perform a pull-in trip to return to the depot from $en(i)$;
6. the single *return arc* (O^+, O^-), with a "large" cost representing the cost of using one more vehicle, which allows any number of units of flow, i.e., vehicles, to depart from O^- and reach O^+.

All the arcs have capacity 1 save the return one, which has capacity equal to the maximum fleet cardinality (if any, $+\infty$ otherwise), and all nodes have deficit 0 (a *circulation* problem). Although this formulation is reminiscent of the classical flow formulation of single-depot VS problems [1] that are solved in the classical planning sequence after the TT stage, it is different because the latter only contains (pairs of) nodes corresponding to the trips of some *input* (fixed) timetable π^*, whereas in our case the optimal timetable π^* is unknown (being part of the integrated decision), hence G^{VS} has to represent *all* possible trips. Thus, G^{VS} can be a "large" graph, and it is also rather dense because "early" trips are typically compatible with all "late" ones. However, different formulations are possible, e.g. where the depot nodes are treated in a way reminiscent of time-expanded networks, with different trade-offs between number of nodes and arcs, as described in [3].

Integrated Model

The integrated model combines node-arc formulations of the VS graph and the TT graphs for all directions $d \in \mathscr{D}$, where arc flow variables represent pairs of compatible trips in either the TT or the VS:

$$\min \alpha c x + \sum_{d \in \mathscr{D}} c^d y^d \qquad (1)$$

$$\sum_{(m,n) \in A_d^{TT}} y^d_{m,n} - \sum_{(n,m) \in A_d^{TT}} y^d_{n,m} = b^d_n \qquad n \in N_d^{TT}\ ,\ d \in \mathscr{D} \qquad (2)$$

$$y^d_{n,m} \in \{0, 1\} \qquad (n, m) \in A_d^{TT}\ ,\ d \in \mathscr{D} \qquad (3)$$

$$\sum_{(m,n) \in A^{VS}} x_{m,n} - \sum_{(n,m) \in A^{VS}} x_{n,m} = 0 \qquad n \in N^{VS} \qquad (4)$$

$$0 \le x_{n,m} \le u_{n,m} \qquad (n, m) \in A^{VS} \qquad (5)$$

$$\sum_{(n,m) \in B(i)} y^{d(i)}_{n,m} = x_{i^-,i^+} \qquad i \in \mathscr{T} \qquad (6)$$

The MILP consists of four blocks. Constraints (2) are the flow conservation constraints of the two TT blocks, where $y^d_{n,m}$ are the arc flow variables on A_d^{TT}; the deficits b^d_n for TT are all 0 except for $n \in \{O_d^-, O_d^+\}$, which, together with (3), ensures that any y^d solution describes a path between O_d^- and O_d^+ in G_d^{TT} (a TT π_d). Constraints (4) are the flow conservation constraints describing a circulation in G^{VS} (a VS Ω) via the arc flow variables $x_{n,m}$, whose capacities $u_{n,m}$ are all 1 except that of the return arc (O^+, O^-). Finally, (6) are the *linking constraints* ensuring that a trip is performed in the VS if and only if it is chosen by the corresponding TT; $B(i)$ is the backward star of the node of $N_{d(i)}^{TT}$ corresponding to trip i, i.e., the set of all arcs in $A_{d(i)}^{TT}$ entering it. The bi-objective nature of ITTVS is modeled using the *weighted objective function* (1) representing the decision-maker priority between the two objectives. The experience of M.A.I.O.R. was crucial in properly setting the weight α and defining the components of the cost vector c for the VS sub-problem.

Solution Approach

Real-life MILP instances of ITTVS are too hard to be solved directly using a general-purpose solver, therefore a heuristic approach is needed. The *matheuristic* approach currently used in `TTD.XT` is a diving heuristic based on the continuous solution of (1)–(6). At each iteration we fix "some" trips, selecting those corresponding to arc variables with a "large" value in the current continuous solution, and progressively construct a feasible ITTVS solution. The fixing is basically greedy, although a very limited amount of backtracking is allowed when infeasibility of the choices is detected, and the iterative process runs until a complete integer solution is obtained, which means that for all $d \in \mathscr{D}$ the corresponding π_d forms a complete O_d^--O_d^+ path.

The fixing strategy is based on the value of a continuous solution ($\tilde{x}, \tilde{y} = [\tilde{y}^d]_{d \in \mathscr{D}}$). After extensive experiments, the best performing fixing strategy is the so called *V-fix*, as we now detail. At the first iteration, when no trips are fixed, we select the direction $d = \mathbf{AB}$, and we restrict the set of candidates to the trips in the forward star of the corresponding dummy source node O_d^- in the corresponding TT subgraph G_d^{TT}, ordering them in terms of the values of the corresponding $\tilde{y}_i^d$. At the second iteration, with just one trip $\bar{i}$ fixed, we select the opposite direction $d = \mathbf{BA}$ and we restrict the set of candidates to the trips in the forward star of node $\bar{i}^+$ in the VS graph, ordered by the corresponding $\tilde{x}_{(i^-,i^+)}$ variable instead. These two initial fixings form a "V" in the time-space graph used by transport planners to represent a timetable (cf. Fig. 2). In the subsequent iterations, we restrict the set of candidates to the union of the trips in both the forward star and the backward star in the TT graphs of the trips that have been fixed, sorted in descending order of the corresponding $\tilde{y}_i$. We then proceed at fixing the one with highest value, provided that a reasonable balance is kept between the number of trips fixed for each direction.

In order to further reduce the problem size, when we fix a trip i to one, we also fix to zero the trips belonging to the "neighborhood" of i that would violate the TT headway if they were selected, because they are "too close" in time to i. After these fixings, the problem may have become unfeasible, therefore before applying the fixing and proceeding with the next iteration, we check feasibility by solving the corresponding TT and VS subproblems, and in case of failure we backtrack on the decision and move to the next candidate in the list. If the list becomes empty, without any (locally) feasible fixing having been identified, we accept failure and we exit from the heuristic (although this has never happened in practice, after that the fixing rule has been properly tuned).

We also mention that `TTD.XT` implements two different ways to find a continuous solution for (1)–(6): a general-purpose LP solver (in particular, `Clp` [9]) and the (approximate) solution of the *Lagrangian Dual* corresponding to the *Lagrangian relaxation* of the linking constraints (6). The latter approach decomposes ITTVS into three independent sub-problems: two acyclic *shortest-paths* for the TT components, and one *min-cost flow* for the VS component, which can therefore be efficiently solved (also in parallel). The Lagrangian Dual is solved using an implementation of the proximal Bundle approach already used with success in other applications, among which (parallel [2]) approaches to multicommodity flow problems [6, 7]. In this version, the diving heuristic can thereby also be viewed as a *Lagrangian heuristic* [5].

4 Practical Experience

The `TTD.XT` module is one of the main optimization modules of *MTram*, the complete solution for planning and scheduling buses, light rail, streetcar and metro services produced by M.A.I.O.R. *MTram* is a suite of intuitive and integrated software modules that allows public transportation authorities and operators to design the service network and timetable, to optimize vehicle blocks and driver duties, to build customized roster, to calculate optimized running times, and to increase service quality and efficiency. Additionally, *MTram* supports the daily operations management, the paycheck computation, and the company's performance analysis. `TTD.XT` is an intuitive web tool that supports the creation and management of timetables for multiple routes and patterns.

As mentioned in the introduction, the planning process of a public transportation system is usually decomposed into stages, that are solved in sequence. Because of the interdependence of the stages, this inevitably leads to suboptimal solutions, in particular in terms of vehicles and drivers needed, since these are only determined in the later steps of the planning process. The innovative aspect of the `TTD.XT` module is that of integrating the vehicle scheduling phase in the timetabling decisions. The embedded algorithm automatically generates trips that satisfy planners' requirements, such as the number of vehicles, the number of passengers, and the desired headways, while considering dwell times, layover/recovery times, different running times throughout the day, and different headways to effectively create timetables that respect different service levels (e.g. on-peak hours, and low-peak hours services). `TTD.XT` provides an integrated interface for timetable designers to input the required data (one screenshot of which is shown in Fig. 1), view the solutions suggested by the algorithm and possibly modify them (a screenshot of which is shown in Fig. 2).

The `TTD.XT` module can be used whenever a timetable service needs to be planned or re-planned. Usually, "seasonal planning" takes place twice a year, when the timetable changes. Yet, within each season, some adjustments may be required. The updating frequency depends on several factors, that in turn depend on the company policies, but roughly one of the timetable lines needs to be updated at least once a week. Therefore, the tool is used regularly throughout the year.

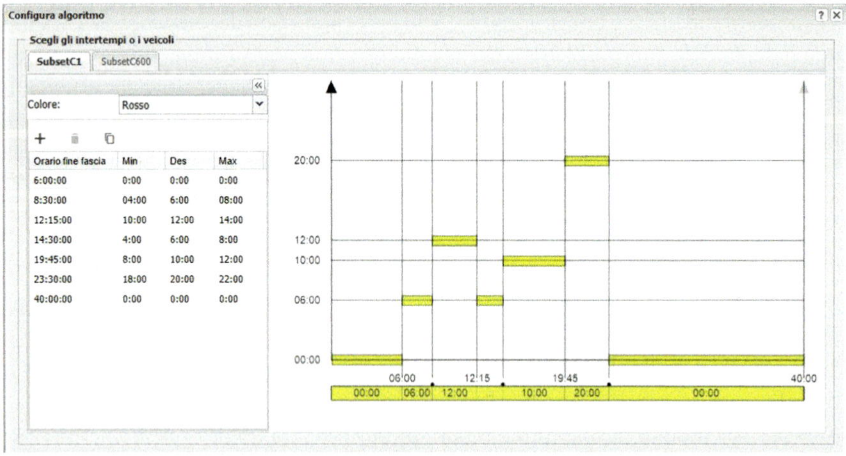

Fig. 1 A screenshot of the `TTD.XT` input interface

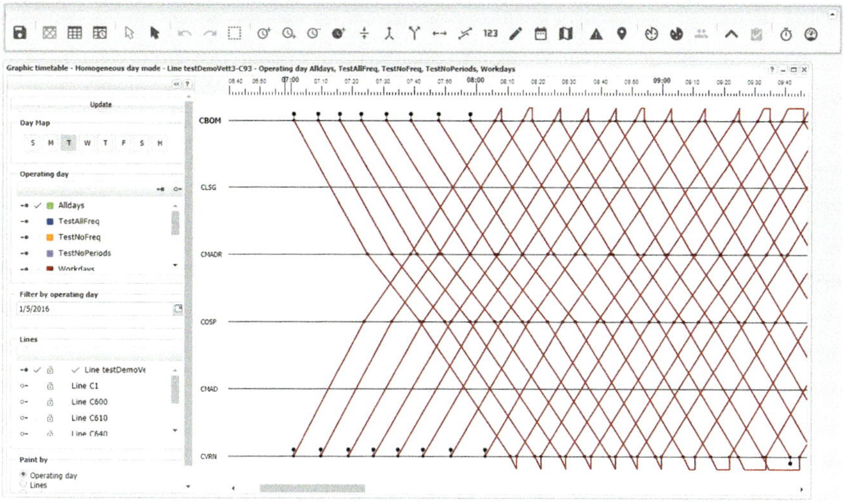

Fig. 2 A screenshot of the `TTD.XT` output interface

As a sample of `TTD.XT` performances we report results on 12 *real-world* bus lines covering the city center of 3 major Italian cities, among which Milano. The data was provided by the corresponding bus service providers, all M.A.I.O.R. customers. Of the 12 instances, 5 correspond to "complex" lines that have more than one *route* for the same direction, which requires some ad-hoc extension of the previously presented models [3]. For each minute of the time horizon there is a possible trip i in $\mathscr{T}$ for each route; since a typical time-horizon ranges roughly from 5:00 to 24:00, there can be as many as 1140 possible trips for each route.

In Table 1 we report some data about the behaviour of the two versions of the heuristic approach: "h-B" using the `Bundle` method to (approximately) solve the Lagrangian dual, and "h-C" using `Clp` (in particular, the version included in `Cbc-2.9.8`, applying the *barrier* at the first iteration and the *dual simplex* in the subsequent ones) to solve the continuous relaxation of (1)–(6). All the experiments have been performed on a PC with a 1.9 GHz Intel Xeon (R) E5-2420 processor. The instances were also tackled by directly applying the commercial MILP solver `Cplex 12.7` with a time limit. All the solvers were finely tuned to find a good trade-off between solution quality and running time; in particular, the stopping criterion used for the `Bundle` method is a maximum number of iterations, and for all our test instances it never reached the optimal value.

Table 1 compares the value of the feasible solution computed by the *fastest* between h-B and h-C (which is not necessarily the one obtaining the best solution) on each specific instance against: (i) the solution constructed manually by the service provider planners, and (ii) the best solution found by `Cplex` on the full MILP formulation (1)–(6) in "comparable" time. To improve readability, we report the *relative gap in percentage* between two values X versus Y, i.e., $(X - Y)/Y \times 100$. In

Table 1 Fastest TTD.XT variant versus manual and Cplex solutions

Instance	B/C	Solution quality: TTD.XT versus			
		Manual	Cplex*1	Cplex*2	Cplex*4
2Cap_R1	h-C	117.93	–	–	4.60
2Cap_R2	h-C	28.84	1863.83	0.75	0.75
2Cap_R3	h-C	35.35	1488.69	1488.69	0.71
2Cap_F1	h-C	202.99	3050.46	2.06	−1.50
2Cap_F2	h-C	51.78	1102.97	885.69	25.72
2Cap_F3	h-C	158.86	5338.87	908.09	−16.23
2Cap_F4	h-C	19.04	–	−7.88	−7.88
3Cap_F5	h-C	252.98	–	48.84	48.61
3Cap_F6	h-C	81.38	–	19.61	19.61
3Cap_F7	h-B	63.53	–	–	–
3Cap_M1	h-B	16.32	–	–	6.11
4Cap_F8	h-B	61.55	–	–	–

particular, the columns "Cplex*k" report the relative gap in percentage between the value of the solution found by Cplex, with a time limit set to k times the running time of (the corresponding variant of) TTD.XT, versus that found by our approach; an entry "–" means that Cplex was not able to find any feasible solution within the time limit. Column "Manual" reports the same measure for the manually constructed solutions versus that found by our approach.

The table shows that TTD.XT finds much better solutions than the manually obtained ones. It has to be remarked that the objective function value is influenced by the choice of the scaling constant α, and therefore it is not trivial to assess how much better the solutions are in practice. Therefore, all the solutions were also inspected by experts of the field, exploiting the previously discussed interface, who have confirmed that they are indeed significantly better than the manual ones. In particular, it can be seen that the VS cost (representing the service operator cost) is generally lower. This can be put down to saving up to three vehicles, which is indeed quite remarkable, considering that fixed vehicle costs are "high" and the total number of required vehicles ranges between 4 and 24. However, interestingly, the main component in the savings was due to a reduction in the waiting times at the terminals; this is actually partly a consequence of the reduction in vehicles (the less the vehicles used, the less the waiting time), but in one case the number of vehicles increased (from 3 to 4), and yet the VS cost significantly decreased. On 3 instances out of 12, more than 80% of the improvement was due to the VS cost. On 4 other instances the two components were more balanced, with a VS improvement ranging from 40 to 60%. On the remaining instances, the TT component contributed to more than 70% of the total improvement.

For smaller instances, and allowing much longer times, CPLEX sometimes finds better solutions (negative figures). In some cases the gap is significant, but in general the heuristic approach is quite competitive, especially considering the many instances in which `Cplex` could not find any feasible solution within the time limit. In particular, for complex lines, a direct solution of the MILP formulation, even with a state-of-the-art MILP solver like `Cplex`, is never competitive. For this cases, the `Bundle` approach appears to be particularly competitive.

The results illustrate the general behaviour of the `TTD.XT` module on real-world ITTVS instances. The module has shown to reliably and consistently improve on the solutions obtained manually, thereby allowing even experienced planners to avoid repetitive and manually-intensive labor and to rather concentrate on more creative tasks, which is the ultimate aim of Decision-Support Systems. Although the integrated model is rather hard to solve by even commercial, state-of-the-art, general-purpose solvers like `Cplex`, the heuristic implemented into `TTD.XT` is quite effective. It is also important to mention that the model is also quite flexible and capable to incorporate other requirements suggested by M.A.I.O.R. customers. Amongst these, we mention *flow control* (i.e., the possibility of fixing the number of vehicles at different points of the planning horizon), and a recently added feature that allows to add specific constraints on the VS to ease the construction of feasible crew duties in the last planning phase, thereby representing a step towards a further level integration that includes timetabling, vehicle scheduling and crew scheduling.

Acknowledgements The authors are very grateful to the bus service operators, among which *ATM* Milan, for providing the instances used for the case studies. This work has received funding from the Europeans Union's EU Framework Programme for Research and Innovation Horizon 2020 under the Marie Skłodowska-Curie Actions Grant Agreement No 764759 "MINOA—Mixed-Integer Non Linear Optimisation: Algorithms and Applications".

References

1. Bertossi, A.A., Carraresi, P., Gallo, G.: On some matching problems arising in vehicle scheduling models. Networks **17**, 271–281 (1987)
2. Cappanera, P., Frangioni, A.: Symmetric and asymmetric parallelization of a cost-decomposition algorithm for multi-commodity flow problems. INFORMS J. Comput. **15**(4), 369–384 (2003)
3. Carosi, S., Frangioni, A., Galli, L., Girardi, L., Vallese, G.: A matheuristic for integrated timetabling and vehicle scheduling. Technical Report, Dipartimento di Informatica, Università di Pisa (2018)
4. Fonseca, J.P., van der Hurk, E., Roberti, R., Larsen, A.: A matheuristic for transfer synchronization through integrated timetabling and vehicle scheduling. Transp. Res. Part B: Methodol. **109**, 128–149 (2018)
5. Frangioni, A.: About Lagrangian methods in integer optimization. Ann. Oper. Res. **139**(1), 163–193 (2005)
6. Frangioni, A., Gallo, G.: A bundle type dual-ascent approach to linear multicommodity min cost flow problems. INFORMS J. Comput. **11**(4), 370–393 (1999)
7. Frangioni, A., Gorgone, E.: Generalized bundle methods for sum-functions with "easy" components: applications to multicommodity network design. Math. Progr. **145**(1), 133–161 (2014)
8. Schmid, V., Ehmke, J.F.: Integrated timetabling and vehicle scheduling with balanced departure times. OR Spectr. **37**(4), 903–928 (2015)
9. The Coin-OR Linear Programming Project. https://projects.coin-or.org/clp

Printed by Printforce, the Netherlands